Springer Proceedings in Mathematics & Statistics

Volume 142

Springer Proceedings in Mathematics & Statistics

This book series features volumes composed of selected contributions from workshops and conferences in all areas of current research in mathematics and statistics, including operation research and optimization. In addition to an overall evaluation of the interest, scientific quality, and timeliness of each proposal at the hands of the publisher, individual contributions are all refereed to the high quality standards of leading journals in the field. Thus, this series provides the research community with well-edited, authoritative reports on developments in the most exciting areas of mathematical and statistical research today.

More information about this series at http://www.springer.com/series/10533

P.G. Romeo · John C. Meakin
A.R. Rajan
Editors

Semigroups, Algebras and Operator Theory

Kochi, India, February 2014

Springer

Editors
P.G. Romeo
Department of Mathematics
Cochin University of Science and Technology (CUSAT)
Kochi, Kerala
India

A.R. Rajan
Department of Mathematics
University of Kerala
Thiruvananthapuram, Kerala
India

John C. Meakin
Department of Mathematics
University of Nebraska
Lincoln, NE
USA

ISSN 2194-1009 ISSN 2194-1017 (electronic)
Springer Proceedings in Mathematics & Statistics
ISBN 978-81-322-2487-7 ISBN 978-81-322-2488-4 (eBook)
DOI 10.1007/978-81-322-2488-4

Library of Congress Control Number: 2015941881

Springer New Delhi Heidelberg New York Dordrecht London

Printed on acid-free paper

Springer (India) Pvt. Ltd. is part of Springer Science+Business Media (www.springer.com)

Preface

The International Conference on Semigroups, Algebras and Operator Theory (ICSAOT-2014) focused on the recent advances in semigroup theory and operator theory. The scientific programs emphasized on the research related to the structure theory of semigroup, semigroup approach to Von Neumann regularity and operator algebras. The conference was hosted by the Department of Mathematics, Cochin University of Science and Technology (CUSAT), Kochi, Kerala, India during 26–28 February 2014.

Leading researchers from 11 different countries working in these areas were invited to participate. A total of 63 delegates including Ph.D. students from eight countries participated. The following plenary lectures were given:

1. John C. Meakin, University of Nebraska, USA—*Amalgams of Inverse Semigroups of C^*—Algebras.*
2. Laszlo Marki, Budapest, Hungary—*Commutative Order in Semigroups.*
3. M.V. Volkov, University of Ekaterinburg, Russia—*The Finite Basis Problem for Kauffman Monoids.*
4. Alessandra Cherubini, Milano, Italy—*Word Problem in Amalgams of Inverse Semigroups.*
5. K.S.S. Nambooripad, Kerala, India—*von Neumann Algebras and Semigroups.*
6. M.K. Sen, Kolkata, India—*Left Clifford Semigroups.*
7. Pascal Weil, LaBRI-CNBS, Talence Cedex, France—*Logic, Language and Semigroups: from the lattice of band varites to the quantifier alteration hierarchy within the 2-variable fragment of first order logic.*
8. Jorge Almeida, Alegre, Portugal—*Irreducibility of Psedovarieties of Semigroup.*
9. B.V. Limaye, Mumbai, India—*Operator Approximations.*
10. S.H. Kulkarni, Chennai, India—*The Null Space Theorem.*
11. M. Thamban Nair, Chennai, India—*Role of Hilbert Scales in Regularization Theory.*

The daily program consisted of lectures, paper presentations and discussions held in an open and encouraging atmosphere.

In addition to the above speakers there were invited talks and paper presentations. We are greatful to all participants for their valuable contributions and for making the ICSAOT-2014 a successfull event. Moreover, we would like to thank the National Board for Higher Mathematics, DAE, Mumbai; CSIR, New Delhi; SERB-DST, New Delhi; KSCSTE, Thiruvananthapuram for providing us with financial support. We are also thankful to the Cochin University of Science and Technology for additional support and practical assistance related to the preparation and organization of the conference. We thank all our referees for their sincere cooperation, which enabled the successful completion of the refereeing processes.

P.G. Romeo
John C. Meakin
A.R. Rajan

Contents

Editors and Contributors

About the Editors

P.G. Romeo is professor and head of the Department of Mathematics at the Cochin University of Science and Technology (CUSAT), Kochi, Kerala, India. His major research interests include semigroups theory, representation theory, universal algebras and category theory. He earned his Ph.D. from the University of Kerala in 1993 under the guidance of Prof. K.S.S. Nambooripad. During his illustrious career as researcher and professor for over two decades, Prof. Romeo taught a wide variety of courses both at undergraduate and graduate levels. He also guides students for doctoral research. He has published research articles with various peer-reviewed international mathematics journals as well as given invited talks at various international and national conferences.

John C. Meakin is the Milton Mohr Professor of Mathematics at the University of Nebraska-Lincoln, USA. He served as the chair of the Department of Mathematics at Nebraska during 2003–2011. His primary research interests include modern algebra and theoretical computer science, with particular focus on the algebraic theory of semigroups and geometric group theory. He teaches a wide variety of courses ranging from freshman-level calculus to advanced graduate courses in algebra and topology. Professor Meakin has published extensively in peer-reviewed international mathematics journals and has served for 20 years as managing editor of *International Journal of Algebra and Computation*. Elected in 2014 as a Fellow of the American Mathematical Society, Prof. Meakin has given around 200 invited lectures at conferences and universities around the world and has held extended visiting positions in Australia, Belgium, France, India, Israel, Italy, Spain and the USA.

A.R. Rajan is an emeritus professor at the Department of Mathematics, University of Kerala, India. He earned his Ph.D. and M.A. (Russian language) from Kerala University. He also held positions such as member of the senate and syndicate of the University of Kerala and chairman of the board of studies in mathematics.

Professor Rajan has also served as member of board of studies at Cochin University of Science and Technology, University of Calicut, Mahatma Gandhi University, Manonmaiam Sundernar University and Periyar University. He has participated in conferences held in Austria, Hungary, United Kingdom, Thailand and Vietnam and published in peer-reviewed journals such as *Semigroup Forum, Quarterly Journal of Mathematics, Journal of Pure* and *Applied Mathematics*. His areas of research interests include structure theory of semigroups, matrix semigroups, topological semigroups, theory of semirings and automata theory.

Contributors

Asma Ali Department of Mathematics, Aligarh Muslim University, Aligarh, India

Farhat Ali Department of Mathematics, Aligarh Muslim University, Aligarh, India

M.S. Balasubramani Department of Mathematics, University of Calicut, Calicut, Kerala, India

Alessandra Cherubini Dipartimento di Matematica, Politecnico di Milano, Milano, Italy

Clauss Haetinger Center of Exact and Technological Sciences, Univates University Centre, Lajeado–RS, Brazil

K.V. Krishna Department of Mathematics, Indian Institute of Technology Guwahati, Guwahati, India

E. Krishnan University College, Thiruvananthapuram, Kerala, India

S.H. Kulkarni Department of Mathematics, Indian Institute of Technology Madras, Chennai, India

Jitender Kumar Department of Mathematics, Indian Institute of Technology Guwahati, Guwahati, India

K. Kumar Department of Mathematics, PSMO College, Tirurangadi, Kerala, India

Balmohan V. Limaye Department of Mathematics, Indian Institute of Technology Bombay, Mumbai, India

Phool Miyan Department of Mathematics, Aligarh Muslim University, Aligarh, India

Rajlaxmi Mukherjee Pal Department of Mathematics, Jadavpur University, Kolkata, India

M.T. Nair Department of Mathematics, IIT Madras, Chennai, India

K.S.S. Nambooripad Department of Mathematics, University of Kerala, Kerala, India

Pavel Pal Department of Mathematics, Jadavpur University, Kolkata, India

A.R. Rajan Department of Mathematics, University of Kerala, Thiruvananthapuram, Kerala, India

Emanuele Rodaro Centro de Matemática, Universidade do Porto, Porto, Portugal

P.G. Romeo Department of Mathematics, Cochin University of Science and Technology, Kochi, Kerala, India

Sujit Kumar Sardar Department of Mathematics, Jadavpur University, Kolkata, India

M.K. Sen Department of Pure Mathematics, University of Kolkata, Kolkata, India

V. Sherly Government College for Women, Thiruvananthapuram, Kerala, India

A.K. Vijayarajan Kerala School of Mathematics, Kozhikode (DT), Kerala, India

P. Vinod Kumar Department of Mathematics, Thunchan Memorial Government College, Tirur, Kerala, India

M.V. Volkov Institute of Mathematics and Computer Science, Ural Federal University, Ekaterinburg, Russia

Decidability Versus Undecidability of the Word Problem in Amalgams of Inverse Semigroups

Alessandra Cherubini and Emanuele Rodaro

Abstract This paper is a survey of some recent results on the word problem for amalgams of inverse semigroups. Some decidability results for special types of amalgams are summarized pointing out where and how the conditions posed on amalgams are used to guarantee the decidability of the word problem. Then a recent result on undecidability is shortly illustrated to show how small is the room between decidability and undecidability of the word problem in amalgams of inverse semigroups.

Keywords Inverse semigroup · Amalgams · Schutzenberger automata · Cactrod inverse automata

1 Introduction

A semigroup S is a regular semigroup if for each $a \in S$ there exists some element $b \in S$ such that $a = aba$ and $b = bab$. Such an element b is called an inverse of a. A regular semigroup where each element has a unique inverse is an inverse semigroup, in such case the (unique) inverse of a is denoted by a^{-1}. Equivalently, an inverse semigroup is a regular semigroup whose idempotents commute, hence its set of idempotents $E(S)$ is a commutative subsemigroup of S, usually called the semilattice of idempotents. A natural partial order is defined on an inverse semigroup S by putting $a \leq b$ if and only if $a = eb$ for some $e \in E(S)$.

Inverse semigroups may be regarded as semigroups of partial one-to-one transformations, so they arise very naturally in several areas of mathematics. We refer the reader to the book of Petrich [25] for basic results and notation about inverse

A. Cherubini (✉)
Dipartimento di Matematica, Politecnico di Milano, Piazza L. da Vinci 32, 20133 Milano, Italy
e-mail: alessandra.cherubini@polimi.it

E. Rodaro
Centro de Matemática, Universidade do Porto, Rua Do Campo Alegre, 687, 4169-007 Porto, Portugal
e-mail: emanuele.rodaro@fc.up.pt

P.G. Romeo et al. (eds.), *Semigroups, Algebras and Operator Theory*, Springer Proceedings in Mathematics & Statistics 142, DOI 10.1007/978-81-322-2488-4_1

semigroups and books of Lawson [19] and Paterson [24] for many references to the connections between inverse semigroups and other branches of mathematics. More recently, inverse semigroups have also attracted the attention of physics and computer science scholars because several notions and tools of inverse semigroup theory are attuned toward questions in solid-state physics (particularly those concerning quasicrystals) and the concrete modeling of time-sensitive interactive systems (see for instance [14, 16, 17]). Besides, inverse semigroup theory gives raise to interesting algorithmic problems.

Inverse semigroups form a variety of algebras defined by the associativity and the following identities:

$$(a^{-1})^{-1} = a, \quad (ab)^{-1} = b^{-1}a^{-1} \qquad (1)$$
$$aa^{-1}a = a, \quad aa^{-1}bb^{-1} = bb^{-1}aa^{-1}$$

Hence, for any given set X, the free inverse semigroup $FIS(X)$ exists, and $FIS(X) = (X \cup X^{-1})^{+}/\nu$ where ν is the congruence generated by the previous identities (Vagner's congruence). Let $T \subseteq (X \cup X^{-1})^{+} \times (X \cup X^{-1})^{+}$. The quotient of the free semigroup $(X \cup X^{-1})^{+}$ by the least congruence τ that contains ν and the relations in T is an inverse semigroup denoted by $S = Inv\langle X; T\rangle$, $\langle X|T\rangle$ is called a presentation of S with set of generators X and T. If both X and T are finite then S is called finitely presented.

The aim of combinatorial inverse semigroup theory is to extract information on an inverse semigroup starting from its presentation, and one of its core problems is the *word problem*: for a given presentation $\langle X|T\rangle$ of an inverse semigroup S

Input: Two words $w, w' \in (X \cup X^{-1})^{+}$
Output: Do w, w' represent the same element of S?

If there exists an algorithm answering to the above question, then S has decidable word problem.

The word problem is, in general, undecidable for inverse semigroups (since Novikov in the early 1950s proved that it is undecidable for groups [23]).

In 1974, Munn gave a nice characterization of the free inverse semigroup $FIS(X)$ in terms of birooted trees labeled on X (i.e., in terms of finite word automata), that gives also a nice solution to the word problem for free inverse semigroups and can be seen as the seed of the theory of presentations of inverse semigroups by generators and relations. The Munn tree $MT(w)$ of a word $w \in (X \cup X^{-1})^{+}$ is the finite subtree of Cayley graph of the free group generated by X obtained by reading the word w starting from 1, with 1 and the reduced word $r(w)$ of w, in the usual group theoretic sense, as initial and final roots. Then the solution of the word problem is the following, two words $w, w' \in (X \cup X^{-1})^{+}$ represent the same element of $FIS(X)$ if and only if they have the same birooted Munn tree.

Munn's work was greatly extended by Stephen [31] who introduced the notion of Schützenberger graphs and Schützenberger automata associated with presentations of inverse semigroups. Let $S = Inv\langle X; T\rangle$, the Schützenberger graph $S\Gamma(X, T; w)$

of the word $w \in (X \cup X^{-1})^+$ associated with the presentation $\langle X|T\rangle$ is the restriction to the vertices that are $\mathcal{R}$-related to w of the Cayley graph of the presentation $\langle X|T\rangle$. The Schützenberger automaton $\mathcal{A}(X, T; w)$ of $w \in (X \cup X^{-1})^+$ associated with the presentation $\langle X|T\rangle$ is the automaton whose underlying graph is $S\Gamma(X, T; w)$ with initial and final vertices $ww^{-1}\tau$ and $w\tau$, respectively. The importance of Schützenberger automata stems from the fact that any two words $w, w' \in (X \cup X^{-1})^+$ represent the same element of $S = Inv\langle X; T\rangle$ if and only if $\mathcal{A}(X, T; w) = \mathcal{A}(X, T; w')$. Therefore, it is clear that an algorithm for determining the Schützenberger graphs of any word associated to a given presentation would solve the word problem for that presentation. In [31] Stephen provides an iterative procedure to construct these automata that, however, is not effective because in general Schützenberger automata are not finite. In spite of that, these automata are widely used in the study of algorithmic and structural questions for several classes of inverse semigroups (see, for instance [3, 5–11, 13, 15, 26–28, 30, 32]).

In this paper, we consider the word problem for amalgamated free products of two given inverse semigroups. The word problem is decidable for amalgamated free products of groups and is undecidable for amalgamated free products of semigroups (even when the two semigroups are finite [29]) but it is not known under which conditions on the inverse semigroups the word problem for amalgamated free products is decidable in the category of inverse semigroups (Problem 5 of [21]). In the sequel, we will briefly illustrate some sufficient conditions on amalgams of inverse semigroups for the word problem being decidable in the amalgamated free products [7, 8] and a negative recent result [28]. The paper is organized as follows. In Sect. 2, we recall basic definitions and relevant results concerning Schützenberger automata of inverse semigroups, and the structure and properties of Schützenberger automata of amalgams of inverse semigroups. In Sect. 3, we briefly describe the constructions to build the Schützenberger automata for some special classes of amalgams of inverse semigroups. In Sect. 4, we provide some sufficient conditions that guarantee these constructions are effective. In Sect. 5, we give a brief description of the techniques to prove that the word problem is undecidable even if we assume nice conditions on the amalgam. The last section is devoted to some conclusions.

2 Preliminaries

In this section, we recall definitions and results concerning Schützenberger automata of inverse semigroups. An *inverse word graph* over an alphabet X is a strongly connected labeled digraph whose edges are labeled over $X \cup X^{-1}$, where X^{-1} is the set of formal inverses of elements in X, so that for each edge labeled by $x \in X$ there is an edge labeled by x^{-1} in the reverse direction. A finite sequence of edges $e_i = (\alpha_i, a_i, \beta_i)$, $1 \le i \le n$, $a_i \in X \cup X^{-1}$ with $\beta_i = \alpha_{i+1}$ for all i with $1 \le i < n$, is an $\alpha_1 - \beta_n$ path of Γ labeled by $a_1 a_2 \ldots a_n \in (X \cup X^{-1})^+$. An *inverse automaton* over X is a triple $\mathcal{A} = (\alpha, \Gamma, \beta)$ where Γ is an inverse word graph over X with set of vertices $V(\Gamma)$, set of edges $Ed(\Gamma)$, and $\alpha, \beta \in V(\Gamma)$ are two special vertices called the initial and final state of $\mathcal{A}$. The language $L[\mathcal{A}]$ recognized by $\mathcal{A}$ is the set of labels of all $\alpha - \beta$ paths of Γ. The inverse word graph Γ over X is *deterministic* if for each $\nu \in V(\Gamma)$, $a \in X \cup X^{-1}$, $(\nu, a, \nu_1), (\nu, a, \nu_2) \in Ed(\Gamma)$ implies $\nu_1 = \nu_2$.

Morphisms between inverse word graphs are graph morphisms that preserve labeling of edges and are referred to as *V-homomorphisms* in [31]. If Γ is an inverse word graph over X and ρ is an equivalence relation on the set of vertices of Γ, the corresponding quotient graph Γ/ρ is called a *V-quotient* of Γ (see [31] for details). There is the least equivalence relation on the vertices of an inverse automaton Γ such that the corresponding V-quotient is deterministic. A deterministic V-quotient of Γ is called a *DV-quotient*. There is a natural homomorphism from Γ onto a V-quotient of Γ. The notions of morphism, V-quotient, and DV-quotient of inverse graphs extend analogously to inverse automata (see [31]).

Let $S = Inv\langle X; T\rangle \simeq (X \cup X^{-1})^+/\tau$ be an inverse semigroup. The Schützenberger graph $S\Gamma(X, T; w)$ of a word $w \in (X \cup X^{-1})^+$ relative to the presentation $\langle X|T\rangle$ is an inverse word graph whose vertices are the elements of the $\mathcal{R}$-class of $w\tau$ in S and whose set of of edges consists of all the triples (s, x, t) with $s, t \in V(S\Gamma(X, T; w)), x \in X \cup X^{-1}$, and $s(x\tau) = t$. We view the edge (s, x, t) as being directed from s to t. The graph $S\Gamma(X, T; w)$ is a deterministic inverse word graph over X. The automaton $\mathcal{A}(X, T; w)$, whose underlying graph is $S\Gamma(X, T; w)$ with $ww^{-1}\tau$ as initial state and $w\tau$ as terminal state, is called the Schützenberger automaton of $w \in (X \cup X^{-1})^+$ relative to the presentation $\langle X|T\rangle$.

In [31] Stephen provides an iterative (but in general not effective) procedure to build $\mathcal{A}(X, T; w)$ via two operations, *the elementary determination* and *the elementary expansion*. We briefly recall such operations. Let Γ be an inverse word graph over X, an elementary determination consists of folding two edges starting from the same vertex and labeled by the same letter of the alphabet $X \cup X^{-1}$. The elementary expansion applied to Γ relative to a presentation $\langle X|T\rangle$ consists in adding a path (ν_1, r, ν_2) to Γ wherever (ν_1, t, ν_2) is a path in Γ and $(r, t) \in T \cup T^{-1}$.

An inverse word graph is called *closed* relative to the presentation $\langle X|T\rangle$, if it is a deterministic word graph where no expansion relative to $\langle X|T\rangle$ can be performed. An inverse automaton is *closed* relative to $\langle X|T\rangle$ if its underlying graph is closed. We define the closure of an inverse automaton $\mathcal{B}$ relative to a presentation $\langle X|T\rangle$ to be a closed automaton $cl(\mathcal{B})$ relative to $\langle X|T\rangle$, such that $L(\mathcal{B}) \subseteq L(cl(\mathcal{B}))$, and if $\mathcal{C}$ is any other closed automaton relative to $\langle X|T\rangle$ such that $L(\mathcal{B}) \subseteq L(\mathcal{C})$ then $L(cl(\mathcal{B})) \subseteq L(\mathcal{C})$. The existence of a unique automaton with these properties follows from the works of Stephen [31, 32] who proved that any two sequences of expansions and determinations of a finite inverse automaton which terminate in a closed inverse automaton, yield to the same inverse automaton. The linear automaton of the word $w = y_1 y_2 \ldots y_m$ with $y_i \in (X \cup X^{-1})$, $1 \le i \le m$ is the automaton $lin(w)$, whose initial and final states are, respectively, α and β and whose underlying graph has $m + 1$ vertices $\alpha = x_0, x_1, \ldots, x_m = \beta$ and edges (x_{i-1}, y_i, x_i) for $1 \le i \le m$. The Schützenberger automaton $\mathcal{A}(X, T; w)$ of w relative to $\langle X|T\rangle$ is $cl(lin(w))$. An *approximate* automaton of $\mathcal{A}(X, T; w)$ is an inverse automaton $\mathcal{A}_1$, such that $w' \in L[\mathcal{A}_1]$ for some $w' \in (X \cup X^{-1})^+$ with $w'\tau = w\tau$ and $L[\mathcal{A}_1] \subseteq L[\mathcal{A}(X, T; w)]$. Obviously, $lin(w)$ is an approximate automaton of $\mathcal{A}(X, T; w)$. Stephen proved that each expansion or determination applied to an approximate automaton $\mathcal{A}_1$ of $\mathcal{A}(X, T; w)$ gives raise to an approximate automaton $\mathcal{A}_2$ which better approximates $\mathcal{A}(X, T; w)$, in the sense that

$$L[\mathcal{A}_1] \subseteq L[\mathcal{A}_2] \subseteq L[\mathcal{A}(X, T; w)] = \{v \in (X \cup X^{-1})^+ | v\tau \geq w\tau\}.$$

Then for any $w, w' \in (X \cup X^{-1})^+$ $w\tau = w'\tau$ if and only if they have the same Schützenberger automata relative to the presentation $\langle X | T \rangle$. Hence one can solve the word problem for a presentation of an inverse semigroup S if he is able to effectively construct the associated Schützenberger automaton, or if he is able to provide a "good enough" approximation of it. We remark that Schützenberger automata in case of free inverse semigroups (i.e., semigroups with presentation $\langle X | \emptyset \rangle$) reduce to Munn trees. Hence the Schützenberger automaton of a word relative to $\langle X | T \rangle$ (or its Munn tree in case of free inverse semigroups) can be seen as a "graphical normal form" of that word in $Inv\langle X; T \rangle$. However, in general there is no effective way for "computing" this normal form, and each case must be considered individually. Indeed, for some families of inverse semigroups the confluence of the above procedure allows an ordered sequence of expansions and determinations that brings to a more expressive description of the shape of Schützenberger automata. In these cases, it is possible to effectively construct approximate automata of the Schützenberger automata which are good enough to solve the word problem. One of these cases consists in amalgams of some classes of inverse semigroups which are introduced in the next section.

3 The Schützenberger Automata of Amalgams of Inverse Semigroups

An amalgam of (inverse) semigroups is a tuple $[S_1, S_2; U, \omega_1, \omega_2]$, where S_1, S_2, U are disjoint (inverse) semigroups and $\omega_i : U \hookrightarrow S_i, i = 1, 2$ are two embeddings. The amalgam can be shortly denoted by $[S_1, S_2; U]$. An amalgam $[S_1, S_2; U, \omega_1, \omega_2]$ of semigroups (groups) is said to be strongly embeddable in a semigroup (group) S if there are injective homomorphisms $\phi_i : S_i \to S$ such that $\phi_1|_U = \phi_2|_U$ and $S_1\phi_1 \cap S_2\phi_2 = U\phi_1 = U\phi_2$. It is well known that every amalgam of groups embeds in a group while semigroup amalgams do not necessarily embed in any semigroup [18]. On the other hand, every amalgam of inverse semigroups embeds in an inverse semigroup, and hence in the corresponding amalgamated free product in the category of inverse semigroups [12] (which is defined by the usual universal construction). In this section, we present an ordered procedure for building the Schützenberger graphs (automata) of the amalgamated free products for some amalgams of inverse semigroups. This construction is provided in [8] and is mainly along the lines of the one given in [2] for the amalgamated free products of a class of amalgams which satisfies some quite technical conditions are recalled in the following definition.

Definition 1 An amalgam $[S_1, S_2; U]$ in the category of inverse semigroups is *lower bounded* if satisfies the following conditions

1. For $i \in \{1, 2\}$ and for all $e \in E(S_i)$ the set $U_i(e) = \{u \in U \mid e \leq_i u\}$, where $\leq_i$ denotes the natural order of S_i, is either empty or contains a least element denoted $f_i(e)$.
2. For $i \in \{1, 2\}$, if $e_1U,\ e_2U, \ldots$ is a descending chain of left cosets in S_i, where $U_i(e_k) \neq \emptyset$ for all $k > 0$, then there exists a positive integer N such that for each $g \in E(U)$ with $g \leq f(e_N)$, $f(e_N \cdot g) = g$.

To describe this construction, we recall some terminology from [2, 8].

If $(\alpha_1, \Gamma_1, \beta_1)$ and $(\alpha_2, \Gamma_2, \beta_2)$ are inverse word automata, then $(\alpha_1, \Gamma_1, \beta_1) \times (\alpha_2, \Gamma_2, \beta_2)$ is the inverse word automaton $(\alpha_1, \Gamma_3, \beta_2)$ where Γ_3 is the V-quotient of the union of Γ_1 and Γ_2 that identifies β_1 with α_2. Let $S_i = Inv\langle X_i; R_i\rangle = (X_i \cup X_i^{-1})^+/\eta_i, i = 1, 2$, where X_1, X_2 are disjoint alphabets. Let $[S_1, S_2; U, \omega_1, \omega_2]$ be an amalgam, we view the natural image of $u \in U$ in S_i under the embedding ω_i as a word in the alphabet X_i, then $\langle X_1 \cup X_2 | R_1 \cup R_2 \cup W\rangle$ with $W = \{(\omega_1(u), \omega_2(u)) | u \in U\}$ is a presentation of amalgamated free product $S_1 *_U S_2$. We put $X = X_1 \cup X_2$ and $R = R_1 \cup R_2$ and we call $\langle X | R \cup W\rangle$ the *standard presentation of* $S_1 *_U S_2$ *with respect to the presentations of* S_1 and S_2, for short, the standard presentation of $S_1 *_U S_2 \simeq (X \cup X^{-1})^+/\tau$. Each $w \in (X \cup X^{-1})^+$ can be factorized in a unique way as $w = w_{1,1}w_{2,1}w_{1,2}w_{2,2} \ldots w_{1,n}w_{2,n}$ where $n > 0$, $w_{1,1} \in (X_1 \cup X_1^{-1})^*$, $w_{2,n} \in (X_2 \cup X_2^{-1})^*$, $w_{1,i} \in (X_1 \cup X_1^{-1})^+$, $w_{2,i} \in (X_2 \cup X_2^{-1})^+$ for all i with $2 \leq i \leq n-1$. We call the above factorization of w the *chromatic factorization* and $n = ||w||$ the *chromatic length* of w.

Let Γ be an inverse word graph over $X = X_1 \cup X_2$ with $X_1 \cap X_2 = \emptyset$, an edge of Γ that is labeled from $X_i \cup X_i^{-1}$, for some $i \in \{1, 2\}$, is said to be *colored* i. A subgraph of Γ is called *monochromatic* if all its edges have the same color. A *lobe* of Γ is defined to be a maximal monochromatic connected subgraph of Γ. The coloring of edges extends to a coloring of lobes. Two lobes are said to be *adjacent* if they share common vertices, called *intersection vertices*. If $\nu \in V(\Gamma)$ is an intersection vertex, then it is common to two unique lobes, which we denote by $\Delta_1(\nu)$ and $\Delta_2(\nu)$, colored 1 and 2, respectively. We define the *lobe graph* of Γ to be the graph whose vertices are the lobes of Γ and whose edges correspond to the adjacency of lobes.

We remark that a nontrivial inverse word graph Δ colored i and closed relative to $\langle X_i | R_i\rangle$ contains all the paths (ν_1, v', ν_2) with $v' \in (X_i \cup X_i^{-1})^+$ such that $v'\eta_i = v\eta_i$, provided that (ν_1, v, ν_2) is a path of Δ. Hence, we say that there is a path (ν_1, s, ν_2) with $s \in S_i$ in Δ whenever $\{(\nu_1, v, \nu_2) | v\eta_i = s\} \neq \emptyset$. Similarly, we say that (ν_1, u, ν_2) with $u \in U$ is a path of Δ to mean that $(\nu_1, \omega_i(u), \nu_2)$ is a path of Δ. For all $\nu \in V(\Delta)$ we denote by $\mathcal{L}_U(\nu, \Delta)$, the set of all the elements $u \in U$ such that (ν, u, ν) is a loop based at ν in Δ.

We say that an inverse word graph Γ is *cactoid* if its lobe graph is a finite tree and adjacent lobes have precisely one common intersection.

Obviously, the linear automaton of a word w with chromatic factorization $w = w_{1,1}w_{2,1}w_{1,2}w_{2,2} \ldots w_{1,n}w_{2,n}$ is a cactoid automaton with at most $2n$ lobes. We have previously remarked that to get the Schützenberger automaton of w relative to the presentation $\langle X | R \cup W \rangle$ of $S_1 *_U S_2$ we have to apply a sequence of expansions and determinations starting from $lin(w)$ until a closed automaton is reached, and we can perform such sequence of operations in any order. Therefore, starting from $lin(w)$, we apply all the expansions and determinations relative to the relations in R using the following operations grouped in constructions.

Step 1: Let $\mathcal{B} = (\alpha, \Gamma, \beta)$ be a cactoid inverse automaton over X. Let Δ be a lobe of Γ, colored i, that is not closed relative to $\langle X_i | R_i \rangle$. Let λ be any vertex of Δ, let $cl(\Delta)$ be a disjoint copy of the closure of Δ relative to $\langle X_i | R_i \rangle$, and let λ^* denote the natural image of λ in $cl(\Delta)$. Construct the automaton $(\lambda, \Gamma, \lambda) \times (\lambda^*, cl(\Delta), \lambda^*)$ and let Γ' be its underlying graph. Let κ be the least V-equivalence that makes Γ' deterministic. The procedure outputs the automaton $\mathcal{B}' = (\alpha', \Gamma'/\kappa, \beta')$, where α', β' denote the respective images of α and β.

The automaton $\mathcal{B}'$ is also a cactoid inverse automaton with a number of lobes less or equal to the number of lobes of $\mathcal{B}$ and if $\mathcal{B}$ approximates $\mathcal{A}(X, R \cup W; w)$ then so does $\mathcal{B}'$. Moreover $\mathcal{B}'$ is finite if and only if $cl(\Delta)$ and $\mathcal{B}$ are finite. Using the previous elementary step, we can perform the following Construction 1.

Construction 1: Let $\mathcal{B} = (\alpha, \Gamma, \beta)$ be a cactoid inverse automaton. Apply iteratively Step 1 till a cactoid inverse automaton $\mathcal{B}^*$ is obtained such that all its lobes are closed relative to $\langle X_i | R_i \rangle$.

If Step 1 terminates for each lobe, this construction terminates, because the number of lobes does not increase at each application of Step 1. Then if the closure of each lobe is a finite graph, $\mathcal{B}^*$ is a finite cactoid automaton whose lobes are finite DV-quotients of Schützenberger lobes relative to $\langle X_i | R_i \rangle$ for some $i \in \{1, 2\}$. We refer to Lemma 4 of [8] for the proof. Of course since in that paper S_1 and S_2 were assumed finite, the closures of the lobes were always finite automata and the construction always terminates. In general, it is enough to ask that S_1 and S_2 have finite $\mathcal{R}$-classes to guarantee the same result. Moreover, we notice that if the above procedure is applied to $lin(w)$ then $\mathcal{B}^*$ is the Schützenberger automaton of w in the free product $S_1 * S_2$, and its lobes are Schützenberger automata relative to $\langle X_i | R_i \rangle$ for some $i \in \{1, 2\}$. Roughly speaking the above procedure means that we are performing on the starting catoid automaton $\mathcal{B}$ all the possible expansions and determinations relative to the presentation $\langle X | R \rangle$.

Let ν be an intersection vertex of an inverse automaton over X, with corresponding lobes $\Delta_1(\nu)$ and $\Delta_2(\nu)$. Let $e_i(\nu)$ denote the minimum idempotent of S_i labeling a loop based at ν in Δ_i (for $i = 1, 2$) and let $L_i(\nu) = \mathcal{L}_U(\nu, \Delta_i(\nu))$. If $L_i(\nu)$ is nonempty, it is a subsemigroup of U. Assume that it has a minimum idempotent which we denote by $f(e_i)$. This hypothesis is trivially satisfied in amalgams of finite inverse semigroups and it is also satisfied in the lower bounded case as a consequence of condition 1 in the definition. Namely, we remark that if $\Delta_i(\nu)$ is a Schützenberger graph, then $L_i(\nu) = U_i(e_i(\nu))$ (see Definition 1), and the hypothesis that $U_i(e_i(\nu))$ has a minimum simplifies next constructions since no V-quotient is needed.

We say that an inverse automaton $\mathcal{B}$ over X has the *property L* if for each intersection vertex ν either $L_1(\nu), L_2(\nu)$ are both empty or $f(e_1(\nu)) = f(e_2(\nu))$. We say that $\mathcal{B}$ has the *loop equality property* if $L_1(\nu) = L_2(\nu)$ for each intersection vertex ν of $\mathcal{B}$. Obviously, each automaton with the loop equality property has the L property. To obtain a cactoid automaton that satisfies the loop equality property, we have to perform a sequence of suitable expansions using relations in W applied to loops based at intersection vertices and relative determinations. This series of operations is grouped into the following Step 2(a):

Step 2(a): Let $\mathcal{B} = (\alpha, \Gamma, \beta)$ be a cactoid inverse automaton over X whose lobes are closed DV-quotients of Schützenberger automata relative to $\langle X_i | R_i \rangle$ for some $i \in \{1, 2\}$ and suppose that it has not the L property, i.e., for some intersection vertex ν and some $i \in \{1, 2\}$ $L_i(\nu) \neq \emptyset$ and $\omega_{3-i}(f(e_i(\nu))) \notin L_{3-i}(\nu)$. Let $f = \omega_{3-i}(f(e_i(\nu)))$ and form the product $\mathcal{C} = (\nu, \Gamma, \nu) \times \mathcal{A}(X_{3-i}, R_{3-i}; f)$. The union of the images of $\Delta_{3-i}(\nu)$ and $\mathcal{A}(X_{3-i}, R_{3-i}; f)$ is a lobe of $\mathcal{C}$ that is a V-quotient of a Schützenberger automaton relative to R_{3-i} by Lemma 3 of [8]. By applying Construction 1 we obtain a cactoid automaton $\mathcal{C}' = (\nu', \Gamma', \nu')$ which is closed relative to $\langle X | R \rangle$ and whose lobes are closed DV-quotients of Schützenberger automata relative to $\langle X_i | R_i \rangle$ for some $i \in \{1, 2\}$. Let $\mathcal{B}' = (\alpha', \Gamma', \beta')$ (where α' and β' are the images of α and β, respectively) be the automaton obtained from $\mathcal{B}$ by the application of Step 2(a) to $\mathcal{B}$ at the vertex ν.

Step 2(a) is used to perform the following

Construction 2(a): Let $\mathcal{B} = (\alpha, \Gamma, \beta)$ be a cactoid inverse automaton. Iteratively apply Step 2(a) at any intersection vertex till a cactoid inverse automaton $\mathcal{B}^*$ which has the property L is obtained.

If this procedure terminates in finitely many steps and $\mathcal{B}$ is finite, then $\mathcal{B}^*$ is a finite deterministic cactoid inverse automaton $\mathcal{B}^*$ whose lobes are closed DV-quotients of Schützenberger automata relative to either $\langle X_i | R_i \rangle$ for some $i \in \{1, 2\}$ which satisfies the L property. Moreover if $\mathcal{B}$ approximates $\mathcal{A}(X, R \cup W; w)$, then $\mathcal{B}^*$ does the same. We refer the reader to Lemma 2.8 of [2] for the proof that under condition 2 of the lower bounded definition, construction 2(a) always terminates in finitely many steps. Obviously, it terminates also under the hypothesis that S_1 and S_2 are finite. Indeed, there are only finitely many possible graphs that can arise as closed DV-quotients of Schützenberger automata relative to $\langle X_i | R_i \rangle$ for some $i \in \{1, 2\}$. Any application of Step 2(a) at a vertex ν replaces a closed DV-quotient of a Schützenberger graph by another closed DV-quotient of a Schützenberger graph, and the new graph has either more edges or more loops (i.e., has higher rank fundamental group) than the original graph. The finiteness of each S_i puts an upper bound on the number of edges and the rank of the fundamental group of these graphs.

Let $\mathcal{B}$ be any cactoid inverse automaton over X whose lobes are closed DV-quotients of Schützenberger automata relative to $\langle X_i | R_i \rangle$ for some $i \in \{1, 2\}$, and let ν be any intersection vertex of two lobes $\Delta_1(\nu)$ and $\Delta_2(\nu)$ of $\mathcal{B}$ such that $f(e_1(\nu)) = f(e_2(\nu))$. Hence, if $(\nu, \omega_i(u), \nu_i)$ for some $u \in U$ is a path in $\Delta_i(\nu)$ then there is also the path $(\nu, \omega_{3-i}(u), \nu_{3-i})$ in $\Delta_{3-i}(\nu)$ (see Lemma 5 in [8]). The two vertices ν_i and ν_{3-i} are called *related pair*. It may happen that one of these paths is a loop and the other not (i.e., $\mathcal{B}$ does not satisfy the loop equality property). We remark that this situation does not happen if condition 1 of Definition 1 holds, and in general when all the lobes of $\mathcal{B}^*$ are Schützenberger graphs. In order to reach the loop equality property, we perform a series of operations that are described in the following Step 2(b).

Step 2(b): Let $\mathcal{B} = (\alpha, \Gamma, \beta)$ be a cactoid inverse automaton over X whose lobes are closed DV-quotients of Schützenberger automata relative to $\langle X_i | R_i \rangle$ for some $i \in \{1, 2\}$, and suppose that $\mathcal{B}$ has the property L but does not satisfy the loop equality property. Then there exists some intersection vertex ν of $\mathcal{B}$ and a non-idempotent element $u \in U$ such that $\omega_i(u) \in L_i(\nu)$ and $\omega_{3-i}(u) \notin L_{3-i}(\nu)$. Then in $\Delta_{3-i}(\nu)$ there is a $(\nu, \omega_{3-i}(u), \nu')$ path for some $\nu' \neq \nu$. Form the V-quotient Γ' of Γ obtained by identifying ν and ν' in $\Delta_{3-i}(\nu)$. Then apply Construction 1 and Construction 2(a) to the resulting automaton.

To get the loop equality property, we need to iteratively apply Step 2(b), this is described in the following

Construction 2(b): Let $\mathcal{B} = (\alpha, \Gamma, \beta)$ be a cactoid inverse automaton over X whose lobes are closed DV-quotients of Schützenberger automata relative to $\langle X_i | R_i \rangle$ for some $i \in \{1, 2\}$ which satisfies property L. Apply iteratively Step 2(b) to each intersection vertex ν with $L_i(\nu) \neq L_{3-i}(\nu)$ till the loop equality property holds.

Step 2(b) can be seen as an elementary expansion applied to automaton $\mathcal{B}$ by sewing a loop labeled by $\omega_{3-i}(u)$ based at the vertex ν because of the relation $(\omega_1(u), \omega_2(u)) \in W$ followed by the associate finite sequence of determinations and Construction 2(a). If S_1 and S_2 have finite $\mathcal{R}$-classes and U is finite then construction 2(b) terminates after finitely many steps to a finite cactoid inverse automaton $\mathcal{B}^*$ whose lobes are closed DV-quotients of Schützenberger automata relative to $\langle X_i | R_i \rangle$ for $i \in \{1, 2\}$ with the loop equality property. Moreover, $\mathcal{B}^*$ approximates $\mathcal{A}(X, R \cup W; w)$ if $\mathcal{B}$ does. Once more, we remark that Construction 2(b) is not needed for amalgams satisfying condition 1 of Definition 1 since the loop equality property is fulfilled after Construction 2(a).

Let $\mathcal{B}$ be a cactoid inverse automaton over X whose lobes are closed DV-quotients of Schützenberger automata relative to $\langle X_i | R_i \rangle$ for some $i \in \{1, 2\}$ and which satisfies the loop equality property. We say that $\mathcal{B}$ has the *related pair separation property* if for any lobe Δ of $\mathcal{B}$ colored i and for any two distinct intersection vertices ν and ν' of Δ there is no word $u \in U$ such that $\omega_i(u)$ labels a $\nu - \nu'$ path in Δ.

Step 3: Let $\mathcal{B} = (\alpha, \Gamma, \beta)$ be a cactoid automaton whose lobes are closed DV-quotients of Schützenberger automata relative to $\langle X_i | R_i \rangle$ for some $i \in \{1, 2\}$ and which satisfies the loop equality property but does not satisfy the related pair separation property. Let ν_0 and ν_1 be two different intersection vertices of a lobe Δ_i, colored i, such that $(\nu_0, \omega_i(u), \nu_1)$ is a path of Δ_i. Since $\mathcal{B}$ has the loop equality property, there is also a path $(\nu_0, \omega_{3-i}(u), \nu_0')$ in $\Delta_{3-i}(\nu_0)$. Consider the graph $\widetilde{\Gamma}$ obtained by disconnecting Γ at ν_0 and replacing ν_0 with $\nu_0(0)$ and $\nu_0(1)$ in $\Delta_{3-i}(\nu_0)$ and Δ_i, respectively. Denote by T_0 the component of

$\widetilde{\Gamma}$ that contains $\nu_0(0)$ and by T_1 the component that contains $\nu_0(1)$. Now put $\mathcal{C} = (\nu_0', T_0, \nu_0') \times (\nu_1, T_1, \nu_1)$. Clearly all the lobes of $\mathcal{C}$ except at most the lobe colored $3-i$ with intersection vertex $\nu_0' = \nu_1$ are closed DV-quotients of Schützenberger automata relative to $\langle X | R \rangle$. Build the automaton $\mathcal{B}' = (\alpha', \Gamma', \beta')$ whose underlying graph is the same of the underlying graph of $\mathcal{C}$, and α', β' are the natural images of α, β. Thus, if we apply Constructions 1, 2(a) and 2(b) to $\mathcal{B}'$ we get a new cactoid automaton satisfying the loop equality property.

Roughly speaking, Step 3 is a sequence of "guessed" expansions and determinations. Namely, since in Δ_i there is a path $(\nu_0, \omega_i(u), \nu_1)$, then because there is the relation $(\omega_1(u), \omega_2(u)) \in W$ we could draw a path $(\nu_0, \omega_{3-i}(u), \nu_1)$. Hence, by iterated determinations of this path with $(\nu_0, \omega_{3-i}(u), \nu_0')$ in $\Delta_{3-i}(\nu_0)$ the two lobes $\Delta_{3-i}(\nu_0)$ and $\Delta_{3-i}(\nu_1)$ glue in a unique lobe colored $3-i$ which has two intersection vertices with Δ_i. The obtained automaton $\mathcal{B}'$ is clearly not a cactoid, and this would prevent the application of Constructions 1, 2(a) and 2(b). Therefore, to be consistent with the cactoid shape, we bypass this problem introducing the cut and paste operation described in Step 3 which has the advantage of generating an approximate automaton.

Construction 3: Let $\mathcal{B} = (\alpha, \Gamma, \beta)$ be a cactoid inverse automaton over X whose lobes are closed DV-quotients of Schützenberger automata relative to $\langle X_i | R_i \rangle$ for some $i \in \{1, 2\}$ which satisfies the loop equality property. Apply iteratively Step 3 till the related pair separation property is reached for each pair of intersection vertices.

Since each lobe has finitely many intersection vertices and Step 3 does not increase the number of lobes, repeated applications of this step will terminate in a finite number of steps in an automaton $\mathcal{B}^*$ that has the related pair separation property. Moreover, if $\mathcal{B}$ approximates $\mathcal{A}(X, R \cup W; w)$ then also the automaton $\mathcal{B}^*$ approximates $\mathcal{A}(X, R \cup W; w)$ by Lemma 8 of [8].

The next step is applied to automata which are not in general cactoid, in particular we are dealing with automata having more then one intersection vertex between two adjacent lobes.

Step 4: Let $\mathcal{B} = (\alpha, \Gamma, \beta)$ be an inverse word automaton whose lobes are closed DV-quotients of some Schützenberger relative to $\langle X_i | R_i \rangle$ for some $i \in \{1, 2\}$, and which has the loop equality property and the related pair separation property. Then, for each intersection vertex ν and for every vertex $\nu_1 \in V(\Delta_i(\nu))$ for which there is a $\nu - \nu_1$ path in $\Delta_i(\nu)$ labeled by $\omega_i(u)$, for some $u \in U$, there exists a unique vertex $\nu_2 \in V(\Delta_{3-i}(\nu))$ such that $\omega_{3-i}(u)$ labels an $\nu - \nu_2$ path in $\Delta_{3-i}(\nu)$. We call the pair ν_1, ν_2 related pair with respect to $\Delta_i(\nu)$, $\Delta_{3-i}(\nu)$. Moreover, the related pair separation property guarantees that ν_1 and ν_2 are not intersection vertices. Build the automaton $\mathcal{B}' = (\alpha', \Gamma', \beta')$ where Γ' is the quotient graph of Γ with respect to the equivalence relation generated by identifying ν_1 with ν_2.

Step 4 can be seen as an expansion followed by the relative sequence of determinations. Namely, if we have in $\Delta_i(\nu)$ a path $(\nu, \omega_i(u), \nu_i)$ for some $u \in U$, then the expansion relative to the relation $(\omega_1(u), \omega_2(u)) \in W$ produces a path

$(\nu, \omega_{3-i}(u), \nu_i)$. Consequently, by the determination with the path $(\nu, \omega_{3-i}(u), \nu_{3-i})$ in $\Delta_{3-i}(\nu)$, the two vertices ν_i and ν_{3-i} are identified. We say that two adjacent lobes are *assimilated* if each related pair of vertices with respect to these two adjacent lobes are identified. The following construction is defined in order to have all the pairs of adjacent lobes assimilated.

Construction 4: Let $\mathcal{B} = (\alpha, \Gamma, \beta)$ be a cactoid inverse automaton over X whose lobes are closed DV-quotients of Schützenberger automata relative to $\langle X_i | R_i \rangle$ for some $i \in \{1, 2\}$ which satisfies the loop equality property and the related pair separation property. Apply iteratively Step 4 with respect to all the related pairs of vertices for each intersection vertex till all pairs of adjacent lobes are assimilated.

If $\mathcal{B}$ is finite Construction 4 terminates in finitely many steps in a finite deterministic inverse word automaton $\mathcal{B}^*$ which approximates $\mathcal{A}(X, R \cup W; w)$ if $\mathcal{B}$ does. In this case we say that $\mathcal{B}^*$ has the *assimilation property*. It is quite easy to verify that in the new intersection vertices between adjacent lobes the loop equality property is preserved (see Lemma 9 of [8]). Moreover, since the assimilation property does not affect the lobes and their adjacency, it turns out that $\mathcal{B}^*$ is a finite inverse word automaton whose lobes are closed DV-quotients of Schützenberger automata relative to $\langle X_i | R_i \rangle$, $i \in \{1, 2\}$, such that its lobe graph is a tree. Obviously $\mathcal{B}^*$ has the adjacent lobe assimilation property and each pair of intersection vertices between two adjacent lobes are connected by related pair of paths (i.e., paths labeled by u for some $u \in U$). Such automaton (graph) is called in [8] an *opuntoid* automaton (graph).

At the end of Construction 4 the obtained automaton is called the *core automaton* of w and it is denoted by $Core(w)$. Note that this is not the Schützenberger automaton of w, and in general it is not the case that $Core(w) = Core(w')$ if $w\tau = w'\tau$. However, the Schützenberger automaton $\mathcal{A}(X, R \cup W; w)$ is obtained from $Core(w)$ by successive applications of another construction called Construction 5 described below. We remark that opuntoid automata (graphs) were defined also in [2] with the main difference that the lobes were Schützenberger graphs relative to $\langle X_i | R_i \rangle$, $i \in \{1, 2\}$. Indeed, conditions 1 of Definition 1 prevents the performing of DV-quotients.

Construction 5 is applied in presence of special vertices. Let Γ be an opuntoid graph, a vertex ν of Γ which is not an intersection vertex belonging to the (unique) lobe Δ_i, $i \in \{1, 2\}$, is called a *bud* of Γ if $L_i(\nu) \neq \emptyset$. The graph Γ is *complete* if it has no buds: an opuntoid automaton is complete if its underlying graph is complete. Of course the occurrence in Δ_i of a path labeled in U requires the application of other expansions (and relative determinations). This can be done as follows.

Step 5: This step consists of the following two substeps:

- Let $\mathcal{B} = (\alpha, \Gamma, \beta)$ be an opuntoid automaton and let $\nu \in V(\Delta_i)$ be a bud belonging to a lobe Δ_i colored $i \in \{1, 2\}$. Put $f = f(e_i(\nu))$ and let $(x, \Lambda, x) = \mathcal{A}(X_{3-i}, R_{3-i}; f)$. Consider the smallest equivalence relation $\rho \subseteq V(\Lambda) \times V(\Lambda)$ which identifies all the vertices of Λ connected to x by some word of $L_i(\nu)$ and such that Λ/ρ is deterministic. Let $\Delta' = \Lambda/\rho$.

- Consider the automaton $\mathcal{C} = (\nu, \Gamma, \nu) \times (x\rho, \Delta', x\rho)$. For all $u \in U$ such that (ν, u, y) and (ν, u, y') are paths of Δ and Δ', respectively, consider the equivalence relation κ on $V(\mathcal{C})$ that identifies y and y'. Build the automaton $\mathcal{B}' = (\alpha', \Gamma', \beta')$ whose underlying graph is the same of the underlying graph of $\mathcal{C}/\kappa$, and α', β' are the natural images of α, β.

The graph $\mathcal{B}'$ produced by the application of Step 5 has one more lobe then $\mathcal{B}$. Furthermore, $\mathcal{B}$ embeds into $\mathcal{B}'$, and it is an opuntoid automaton that approximates $\mathcal{A}(X, R \cup W; w)$ if $\mathcal{B}$ does. This is obvious in the lower bounded case. Indeed, in the lower bounded case, Step 5 becomes simpler: in this case we have that Δ_i is a Schützenberger graph, and the elements in $L_i(e_i(\nu))$ for each bud ν are greater or equal to $f(e_i(\nu))$ with respect to the natural order defined in U, whence the relation ρ is reduced to the identity. Instead for the finite case to prove that these properties hold requires a quite technical argument contained in Lemma 10 in [8]. Note that this lemma only uses the periodicity of U.

Construction 5: Let $\mathcal{B}$ be an opuntoid automaton with at least a bud. Iteratively apply Step 5 till a complete automaton is obtained.

In general, Construction 5 does not terminate and one obtains a directed system of opuntoid automata $\mathcal{A}_0 = Core(w), \mathcal{A}_1, \ldots \mathcal{A}_N \ldots$ whose direct limit is $\mathcal{A}(X, R \cup W; w)$. Therefore, if Constrictions 1–4 terminates in finitely many steps, and an application of Step 5 to some bud of $\mathcal{A}_i$ produces an automaton where $\mathcal{A}_i$ embeds, then by the above discussion we derive that the Schützenberger graph of an amalgamated free product of inverse semigroups is a complete opuntoid automaton.

4 Some Classes of Amalgams with Decidable Word Problem

When an application of Step 5 to some bud of $\mathcal{A}_i$ produces an opuntoid automaton where $\mathcal{A}_i$ embeds, since Step 5 depends locally on the lobe containing the bud at which the step is applied, the information for building the entire Schützenberger automaton is already contained in *Core*(w). In such case, we can solve the word problem for each class of amalgams of inverse semigroups where the above condition on Construction 5 holds, and where *Core*(w) can be effectively built for each word w. Namely, with the notation used in the previous sections, if $w, w' \in (X \cup X^{-1})^+$, we can consider the following recursive procedure: let $\mathcal{C}_0 = Core(w)$ and let α, β be the initial and final states, respectively; consider the opuntoid automaton $\mathcal{C}_{i+1}$ obtained applying Step 5 to all the buds of $\mathcal{C}_i$. Let $\mathcal{C}_j$ be the smallest opuntoid automaton in the above sequence such for any bud ν and any shortest path (α, u, v), $\|w'\| \leq \|u\|$. We put $(\alpha, Ext_{\|w'\|}(Core(w)), \beta) = \mathcal{C}_j$, and analogously build the opuntoid automaton $(\alpha', Ext_{\|w\|}(Core(w')), \beta')$. Hence, the two words w, w' represent the same element of $S_1 *_U S_2$ if and only if $w' \in L[(\alpha, Ext_{\|w'\|}(Core(w)), \beta)]$ and $w \in L[(\alpha', Ext_{\|w\|}(Core(w')), \beta')]$. If we go through the steps of the previous Sect. 3, one gets the following

Theorem 1 (see also Theorem 1, [7]) *Let $[S_1, S_2; U, \omega_1, \omega_2]$ be an amalgam of inverse semigroups. Then Construction of Core(w) is effective if the following conditions are satisfied.*

1. *Each $\mathcal{R}$-class of S_i (equivalently each Schützenberger graph relative to the presentations $\langle X_i | R_i \rangle$) is finite.*
2. *The membership problem for U in each S_i is decidable.*
3. *For any finite cactoid automaton associated with the given amalgam and for any vertices ν, ν' of a lobe colored i and for each $i = 1, 2$, there is an algorithm to decide whether the label of some $\nu - \nu'$ path is in U. Note that this algorithm is equivalent to check the emptiness of $L_i(e_i(\nu))$.*
4. *For any finite cactoid automaton associated with the given amalgam and for any vertex ν of a lobe colored i, (and each $i = 1, 2$) with $L_i(e_i(\nu)) \neq \emptyset$ the least element $f(e_i(\nu))$ of $L_i(e_i(\nu))$ exists and there is an algorithm to compute it.*
5. *Construction 2(a) of the above procedure must terminate after finitely many applications at all the intersection vertices, and there is an effective bound on the number of applications of the Step 2(a) that need to be applied in order for Construction 2(a) to terminate. More precisely, for any cactoid automaton with n lobes associated with the given amalgam, there is an effectively computable bound $\chi(n)$ such that the sequence of cactoid automata obtained by applying Step 2(a) successively at some intersection vertex of the previously constructed automaton in the sequence will terminate after at most $\chi(n)$ steps in a cactoid automaton which has the L property. The bound $\chi(n)$ and the final automaton constructed depend on n and on the structure of the lobes in the original cactoid automaton.*
6. *Construction 2(b) and Construction (4) of the above procedure must terminate after finitely many applications at all intersection vertices of Steps 2(b) and 4. More precisely, for any cactoid automaton with n lobes associated with the given amalgam, there is an effectively computable bound $\eta(n)$ such that the sequence of cactoid automata obtained by applying Step 2(b) successively at some intersection vertex of the previously constructed automaton in the sequence will terminate after at most $\eta(n)$ steps in a cactoid automaton which has the loop equality property. Similarly for any cactoid automaton with n lobes associated with the given amalgam, there is an effectively computable bound $\kappa(n)$ such that the sequence of cactoid automata obtained by applying Step 4 successively at some intersection vertex of the previously constructed automaton in the sequence will terminate after at most $\kappa(n)$ steps in a cactoid automaton which has the assimilation property.*

By the discussion at the end of previous section, we can deduce the following decidability results.

Theorem 2 (Theorem 3.4 [7]) *The word problem is decidable for any inverse semigroup amalgam of the form $S = FIS(A) *_U FIS(B)$ where U is a finitely generated inverse subsemigroup of $FIS(A)$ and $FIS(B)$.*

Theorem 3 (Theorem 2 [8]) *Let $S = S_1 *_U S_2$ be an amalgamated free product of finite inverse semigroups S_1 and S_2 amalgamating a common inverse subsemigroup U, where $S_i = Inv\langle X_i; R_i\rangle$ are given finite presentations of S_i for $i = 1, 2$. Then the word problem for S is decidable.*

We underline that the previous result is quite in contrast with the general case of amalgams of finite semigroups where the word problem has been proven to be undecidable by Sapir [29].

5 Undecidability of the Word Problem

In the opposite trend with respect to the results shown in the previous section, here we survey a recent result of undecidability of the word problem for amalgams of inverse semigroups with nice algorithmic conditions on the initial semigroups. In particular, we sketch the proof of the following theorem.

Theorem 4 ([28]) *There exists an amalgam $[S_1, S_2; U, \omega_1, \omega_2]$ of inverse semigroups such that:*

1. *S_1 and S_2 have finite $\mathcal{R}$-classes (and therefore solvable word problem);*
2. *U is a free inverse semigroup with zero of finite rank;*
3. *the membership problem of $\omega_i(U)$ is decidable in S_i for $i = 1, 2$;*
4. *ω_1, ω_2 and their inverses are computable functions;*

but for which the following problems are undecidable:

(i) the word problem;
(ii) checking whether or not a given $\mathcal{D}$-class of $S_1 *_U S_2$ is finite;

The proof uses Minsky machines, also called 2-counter machines. In the next subsections, we recall some basic definitions of 2-counter machines, and we finally give an idea of the encoding which allow to reduce the word problem to the halting problem for such machines. Although [28] takes inspiration in the usage of Minsky machines from [20, 29], there are several technical differences, starting from the necessity of considering particular subclasses of 2-counter machines more suitable to deal with inverse semigroups.

5.1 The Amalgam Associated to a 2-counter Machine

A 2-counter machine (for short, $CM(2)$) is a system $\mathcal{M} = (Q, \delta, \iota, f)$ with 2 tapes, Q is the nonempty finite set of internal states, $\iota \in Q$ is the initial state, and $f \in Q$ is the final (halting) state. The alphabet used by the tape is $A = \{\perp, a\}$, where $\perp$ is a blank symbol, and

$$\delta \subseteq (Q \times \{1, \dots, 2\} \times A \times Q) \cup (Q \times \{1, \dots, 2\} \times D \times Q)$$

where $D = \{-, 0, +\}$ and the symbols $-, 0, +$ denote, respectively, a shift on the left, no-shift, and right-shift of a head of the machine. Tapes are one-way (rightward) infinite delimited on the leftmost squares by the blank symbol $\perp$, and all the other squares contain the symbol a. The term counter comes from the fact that the head pointing to a square takes track of an integer, namely the number of a's from the black symbol to the pointed square. Each element of δ is thus a quadruple of one of the two forms: (q, i, s, q'), (q, i, d, q') where $q, q' \in Q$, $i \in \{1, 2\}$, $s \in A$ and $d \in D$. The interpretation of a quadruple of the form (q, i, s, q') is that if the machine $\mathcal{M}$ is in the state q and the ith-head (one for each of the two tapes) is reading the symbol s then the machine changes its state into q'. This instruction is used to test whether the content of a counter is zero (the head is reading the symbol $\perp$) or the head is reading a square with symbol a. This kind of instructions is called *test instructions*. On the other hand, an instruction (q, i, d, q') has the following interpretation: if $\mathcal{M}$ is in the state q then the ith-head of $\mathcal{M}$ shifts one cell to the right ($d = +$), left ($d = -$), or it keeps the same position ($d = 0$), and finally the state is changed to q'. As usual, the evolution of $\mathcal{M}$ can be followed through instantaneous descriptions of the machine. An instantaneous description (for short, ID) of a *CM*(2) $\mathcal{M} = (Q, \delta, \iota, f)$ is a 3-tuple $(q, n_1, n_2) \in Q \times \mathbb{N}^2$. It represents that $\mathcal{M}$ is in state q and the ith-head is in position n_i for $i = 1, 2$, where we assume the position of the head reading the symbol $\perp$ to be 0. The relation $\vdash_{\mathcal{M}}$ on the set of configurations associates to a configuration the one that is obtained by applying the transition map δ. More precisely, if (q, n_1, n_2) is an configuration

$$(q, n_1, n_2) \vdash_{\mathcal{M}} (q', n_1', n_2')$$

if one of the following conditions is satisfied for some $i \in \{1, 2\}$:

- $(q, i, \perp, q') \in \delta$ and $n_i = n_i' = 0$.
- $(q, i, a, q') \in \delta$ and $n_i = n_i' > 0$.
- $(q, i, -, q') \in \delta$ and $n_i - 1 = n_i'$.
- $(q, i, 0, q') \in \delta$ and $n_i = n_i'$.
- $(q, i, +, q') \in \delta$ and $n_i + 1 = n_i'$.

As usual, the reflexive and transitive closure of $\vdash_{\mathcal{M}}$ and its application n times are denoted by $\vdash_{\mathcal{M}}^*$ and $\vdash_{\mathcal{M}}^n$, respectively. The pair (n_1, n_2) is accepted by $\mathcal{M}$ if $(\iota, n_1, n_2) \vdash_{\mathcal{M}}^* (f, n_1', n_2')$ for some pair (n_1', n_1').

Contrary to [20, 29], where 2-counter machines are used in their full generality, to prove Theorem 4 it is needed to deal with a subclass of *CM*(2) called *reversible 2-counter machines*. The reason, which will be more clear later, is due to the fact that inverse semigroups are by their nature reversible being represented by partial one-to-one functions. Roughly speaking, a reversible 2-counter machine is a *CM*(2) such that if there is a computation

$$(q_1, n_1, m_2) \vdash_{\mathcal{M}} (q_2, n_2, m_2) \vdash_{\mathcal{M}} \dots \vdash_{\mathcal{M}} (q_k, n_k, m_k)$$

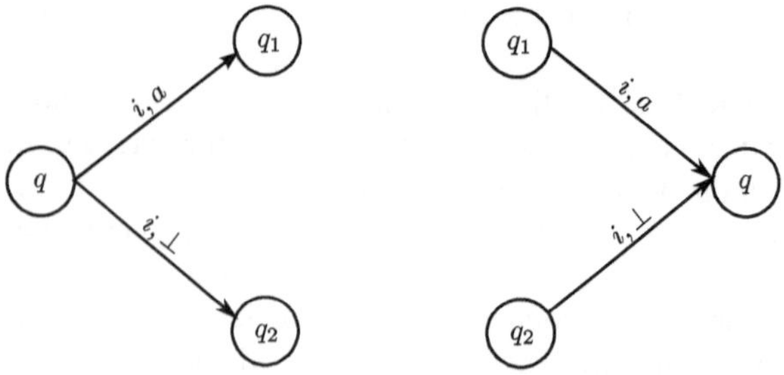

Fig. 1 The deterministic case (on the *left*) and the reversible case (on the *right*)

then this sequence is unique, hence the initial configuration (q_1, n_1, m_2) can be retrieved by (q_k, n_k, m_k). In (theoretical) computer science this notion is central because such machines have no thermodynamical cost (see for instance the seminal paper of Bennet [1]). The following is a more precise definition of reversible deterministic 2-counter machines. Let $\mathcal{M} = (Q, \delta, \iota, f)$ be a *CM*(2), $\mathcal{M} = (Q, \delta, \iota, f)$ can be depicted as a labeled graph $\mathcal{G}(\mathcal{M})$ with set of vertices Q and labeled edges δ where $(q_1, i, h, q_2) \in \delta$ is represented as an arrow from q_1 to q_2 labeled by i, h with $i \in \{1, 2\}$ and $h \in \{a, \perp, +, 0, -\}$. Thus, $\mathcal{M}$ is *deterministic* when the only case where a vertex q of $\mathcal{G}(\mathcal{M})$ may have two outgoing edges is when we have a test instruction (Fig. 1), i.e., $(q, i, a, q_1), (q, i, \perp, q_2)$ are two edges of $\mathcal{G}(\mathcal{M})$ with $i \in \{1, 2\}$. Dually, $\mathcal{M}$ is *reversible* when for each vertex q of $\mathcal{G}(\mathcal{M})$ with in-degree strictly greater than one there are only two ingoing edges of the form $(q_1, i, a, q), (q_2, i, \perp, q)$ for some $i \in \{1, 2\}$ (Fig. 1). It is clear from the definitions that every ID of a deterministic (reversible, respectively) *CM*(2) has at most one ID that immediately follows (precedes) it in some computation. Restricting to these machines does not affect the computational power which remains equivalent to the one of the Turing machines as the following theorem shows.

Theorem 5 ([22]) *For any deterministic Turing machine $\mathcal{T}$ there is a deterministic reversible CM(2) $\mathcal{M}$ that simulates $\mathcal{T}$.*

The strategy to prove Theorem 4 is to encode a general computation $\mathcal{M}\ (\iota, n_1, n_2) \vdash^k_{\mathcal{M}} (q, n_1', n_2')$ in an approximate automaton $\mathcal{B}^k_{n_1,n_2}$ of the Schützenberger automaton of some word w_{n_1,n_2} representing the initial configuration (ι, n_1, n_2) of the machine, and to simulate a one step computation $(q, n_1', n_2') \vdash_{\mathcal{M}} (q', n_1'', n_2'')$ by a suitable expansion. Eventually, in the case the machine reaches the halting state f, a suitable relation forces f to be a zero, thus collapsing the Schützenberger automaton of w_{n_1,n_2} to the Schützenberger automaton of the zero, whence reducing the reachability of the state f to checking whether $w_{n_1,n_2} = 0$. The main technical problem is the control of the expansions. Expansions must be in one-to-one correspondence with each step of the computation, and each of them must have a "local influence". This is done to avoid unexpected quotients, and so loose the information encoded in $\mathcal{B}^k_{n_1,n_2}$. Therefore, in order to fulfill these requirements it becomes clear how determinism and the reversibility of the machine plays a fundamental role. The determinism of $\mathcal{M}$ is required to have uniqueness in the application of each expansion. Indeed, in the nondeterministic case, although the machine can choose nondeterministically for just one transition, all the expansions for all the relations associated to each possible choice have to be performed, and in this case the control of the quotients would be more difficult or even impossible. On the other side, the reversibility is used to avoid "feedback expansions": if one transition can be reached in two (or more) different ways, we should have at least two relations $u_1 = u_2$, $v_1 = v_2$ with $u_2 = v_2$. Now suppose that the expansion corresponding to a one step of the computation is relative to the relation $u_1 = u_2$. This creates a new path labeled by u_2. Since $u_2 = v_2$, this triggers a new expansion relative to the relation $v_1 = v_2$ which is "backward" with respect to the timeline of the computation. Another condition has to be imposed on the machine. The two-counter machine $\mathcal{M} = (Q, \delta, \iota, f)$ is called alternating if, for all pairs of different instructions (q, i, h, q'), $(q', j, h', q'') \in \delta$, it is $j = 3 - i$. The reason why we restrict to the class of alternating machines is purely technical and it is fundamental in simplifying the proof of the finiteness of the $\mathcal{R}$-classes. Although we have restricted to alternating machine this class keeps the same computational power of the Turing machines. Indeed, by adding dummy states it is easy to prove the following proposition.

Proposition 1 *Let* $\mathcal{M} = (Q, \delta, \iota, f)$ *be a deterministic reversible* $CM(2)$*. Then, there is a deterministic reversible and alternating* $CM(2)$ $\mathcal{M}'$ *that simulates* $\mathcal{M}$*.*

A simplification on the general description of the machine can be done: an instruction $(p, i, 0, q)$ can always be replaced by the couple of instructions (p, i, a, q), $(p, i, \perp, q)$ and by doing so the $CM(2)$ remains deterministic, reversible, and alternating. Therefore, a $CM(2)$ which is deterministic, reversible, alternating, and has no instruction of the form $(p, i, 0, q)$ is said to be *normalized.* Taking in particular the universal Turing machine in Theorem 5 and being undecidable whether or not the universal Turing machine can accept a given input, by Proposition 1, the following corollary is obtained.

Corollary 1 *There exists a normalized* $CM(2)$ $\mathcal{M}^*$ *such that it is undecidable whether or not a given* $(m, n) \in \mathbb{N}^2$ *is accepted by* $\mathcal{M}^*$*.*

We now sketch how to associate to $\mathcal{M}^*$ an amalgam for which the word problem is equivalent to the halting problem for $\mathcal{M}^*$. The rough idea is depicted in Fig. 2: the two tapes of the machine $\mathcal{M}$ are encoded by two inverse semigroups S_1, S_2, and the control of $\mathcal{M}$ is handled through a common inverse subsemigroup U.

We start by associating to $\mathcal{M}$ two inverse semigroups S_1, S_2, representing the two tapes, called, respectively, the *left tape inverse semigroup* and the *right tape inverse semigroup* of a normalized 2-counter machine $\mathcal{M} = (Q, \delta, \iota, f)$. The left and right tape inverse semigroups associated to $\mathcal{M}$ are the inverse semigroups S_i $(i = 1, 2)$ presented by $\langle X_i \mid T_i \rangle$, where:

$$X_i = \{\perp_i, a_i, t_i\} \cup \{q_i : q \in Q\},$$

$$T_i = T_i^c \cup T_i^t \cup T_i^w \cup T_i^e \cup T_i^f$$

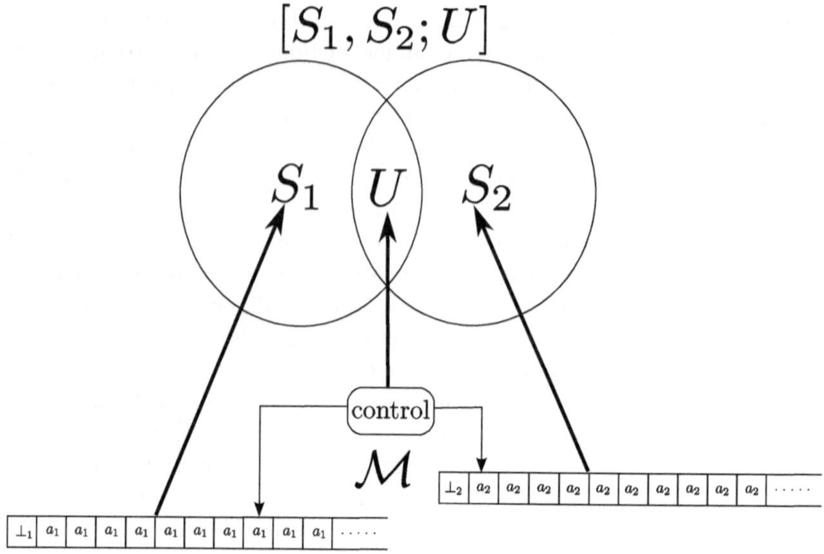

Fig. 2 The rough idea of the encoding

and:

- $\mathcal{T}_i^c$ are the commuting relations, used to keep track of the instantaneous description of the machine;
- $\mathcal{T}_i^t$ are the test relations (corresponding to the test instructions);
- $\mathcal{T}_i^w$ are the writing relations (corresponding to instructions that move the head of the ith tape to the right);
- $\mathcal{T}_i^e$ are the erasing relations (corresponding to instructions that move the head of the ith tape to the left);
- $\mathcal{T}_i^f$ are meant to force the final state f of $\mathcal{M}$ to be a zero and to enforce some finiteness properties on the semigroup S_i.

We do not enter into the details of all the relations involved, instead we will later show the usage of some of them in the simulation of $\mathcal{M}$ via the construction of the Schützenberger automaton. We associate to $\mathcal{M}$ another inverse semigroup U which represents its control unit. The core inverse semigroup U of $\mathcal{M}$ is the free inverse semigroup with zero presented by $\langle X_U | \mathcal{T}_U \rangle$, where $X_U = Q \cup \{t\}$ and the set of relations $\mathcal{T}_U$ which are used to force the final state f to be the zero of the amalgamated free product. The following proposition describes the embeddings ω_i, $i = 1, 2$.

Proposition 2 *Let $\mathcal{M}$ be a normalized 2-counter machine and let S_1, S_2, U be, respectively, the left-right tape inverse semigroups and the core inverse semigroup of $\mathcal{M}$. The map ω_i defined by*

$$\omega_i(t) = t_i, \ \omega_i(q) = q_i \ (q \in Q)$$

can be extended to a monomorphism $\omega_i : U \hookrightarrow S_i$ for $i = 1, 2$. Moreover, the membership problem for $\omega_i(U)$ is decidable in S_i, and both ω_i and its inverse are computable.

In view of Proposition 2, an amalgam can be associated to a normalized 2-counter machine:

Definition 2 Let $\mathcal{M} = (Q, \delta, \iota, f)$ be a normalized 2-counter machine. The amalgam of inverse semigroups associated to $\mathcal{M}$ is the 5-tuple $[S_1, S_2; U, \omega_1, \omega_2]$ where S_1, S_2 are the left-right tape inverse semigroups of $\mathcal{M}$, U is the core inverse semigroup of $\mathcal{M}$ and $\omega_i : U \hookrightarrow S_i$ are the embeddings of Proposition 2. In this way, the amalgamated free product of the amalgam $[S_1, S_2; U, \omega_1, \omega_2]$ associated to $\mathcal{M}$ can be presented by

$$\langle X_1 \cup X_2 \mid \mathcal{T}_1 \cup \mathcal{T}_2 \cup \mathcal{T}_3 \rangle$$

where

$$\mathcal{T}_3 = \{(q_1, q_2) : q \in Q\} \cup \{(t_1, t_2)\}$$

The left-right tape inverse semigroups S_i have the following important property:

Proposition 3 *Let $\mathcal{M}$ be a normalized 2-counter machine and let S_1, S_2 be, respectively, the left-right tape inverse semigroups of $\mathcal{M}$. Then the Green $\mathcal{R}$-classes of S_i are finite for $i = 1, 2$.*

5.2 *Simulating a 2-counter Machines via Stephen's Iterated Procedure*

As we have anticipated in the previous section, we simulate the dynamics of the machine encoding the history of the computation in an approximate automaton and each expansion corresponds to one and only one step of the computation of the machine. Suppose that we have the following computation:

$$(\iota, m, n) = (q^{(0)}, m_0, n_0) \vdash_{\mathcal{M}} \dots \vdash_{\mathcal{M}} (q^{(k)}, m_k, n_k). \tag{2}$$

Since $\mathcal{M}$ is deterministic, there is at most one such sequence of length $k+1$ starting with (ι, m, n). Write $m'_k = \max\{m_0, \dots, m_k\}$, $n'_k = \max\{n_0, \dots, n_k\}$. We associate to the computation (2), a finite inverse X-automaton $\mathcal{B}^{(k)}_{m,n}$ (see Fig. 3) as follows (describing only the edges with positive label):

- The vertices are of the form $c_{i,j}$ and $d_{i,\ell}$ for $i = 0, \dots, k$ and $j = 0, \dots, m'_k + 1$ and $\ell = 0, \dots, n'_k + 1$.
- $c_{0,0}$ is the initial vertex and $d_{0,0}$ is the final vertex.

and there exist the following edges:

- $c_{i-1,j}\xrightarrow{t_1,t_2}c_{i,j}$ for all $i = 1, \ldots, k$ and $j = 0, \ldots, m'_k + 1$.
- $d_{i-1,\ell}\xrightarrow{t_1,t_2}d_{i,\ell}$ for all $i = 1, \ldots, k$ and $\ell = 0, \ldots, n'_k + 1$.
- $c_{i,0}\xrightarrow{\perp_1}c_{i,1}$ for all $i = 0, \ldots, k$.
- $c_{i,j}\xrightarrow{a_1}c_{i,j+1}$ for all $i = 0, \ldots, k$ and $j = 1, \ldots, m'_k$.
- $d_{i,1}\xrightarrow{\perp_2}d_{i,0}$ for all $i = 0, \ldots, k$.
- $d_{i,j+1}\xrightarrow{a_2}d_{i,j}$ for all $i = 0, \ldots, k$ and $j = 1, \ldots, n'_k$.
- $c_{i,m_i+1}\xrightarrow{q_1^{(i)},q_2^{(i)}}d_{i,n_i+1}$ for all $i = 0, \ldots, k$.

In the sequel, a brief justification of how this automaton encodes the computation (2), and how the dynamics of the machine is simulated by Stephen's iterative construction is given. First, note that each ith configuration is encoded in the ith level of the automaton, i.e., the (induced) subgraph with vertices $c_{i,j}, d_{i,\ell}$, with $j = 0, \ldots, m'_k + 1$ and $\ell = 0, \ldots, n'_k + 1$. For instance, the first level encodes the initial configuration $(q^{(0)}, m, n)$, with $q^{(0)} = \iota$, by reading from left to right the path:

$$c_{0,0}\xrightarrow{\perp_1 a_1^m q_1^{(0)} a_2^n \perp_2}d_{0,0}$$

while the second level encodes the configuration $(q^{(1)}, m+1, n)$ by reading the path:

$$c_{1,0}\xrightarrow{\perp_1 a_1^{m+1} q_1^{(1)} a_2^n \perp_2}d_{1,0}$$

Note that each level is separated by the next one with edges labeled by t_1, t_2 that can be interpreted as the unit of time. Furthermore, note that by the definition of the core semigroup $q_1^{(i)} = q_2^{(i)} = q^{(i)}$, and $t_1 = t_2 = t$. Each new configuration corresponding to the transition

$$(q^{(k)}, m_k, n_k) \vdash_{\mathcal{M}} (q^{(k+1)}, m_{k+1}, n_{k+1})$$

is obtained by adding a new $(k+1)$th level separated by the previous one by edges labeled by t's. Let us make a practical example and consider the following computation of an hypothetical normalized $CM(2)$ (Fig. 4):

$$(\iota, 2, 3) \vdash_{\mathcal{M}} (q^{(1)}, 3, 2) \vdash_{\mathcal{M}} (q^{(2)}, 3, 2)$$

where the first transition is due to the instruction $(\iota, 1, +, q^{(1)})$, and the second one is a test instruction $(q^{(1)}, 2, a, q^{(2)})$. Let us start from the linear automaton of the initial configuration represented by the word $w_{2,3} = \perp_1 a_1^2 \iota_1 a_2^3 \perp_2$, and let us start to build the Schützenberger automaton of this word, note that $\iota_1 = \iota_2 = \iota$ by the relations of Definition 2 (see phase 1 of Fig. 4). Inside the set T_1^w, that takes care of all the relations relative to the instructions relative to the first tape, associated to the instruction $(\iota, 1, +, q^{(1)})$ we have the following relations:

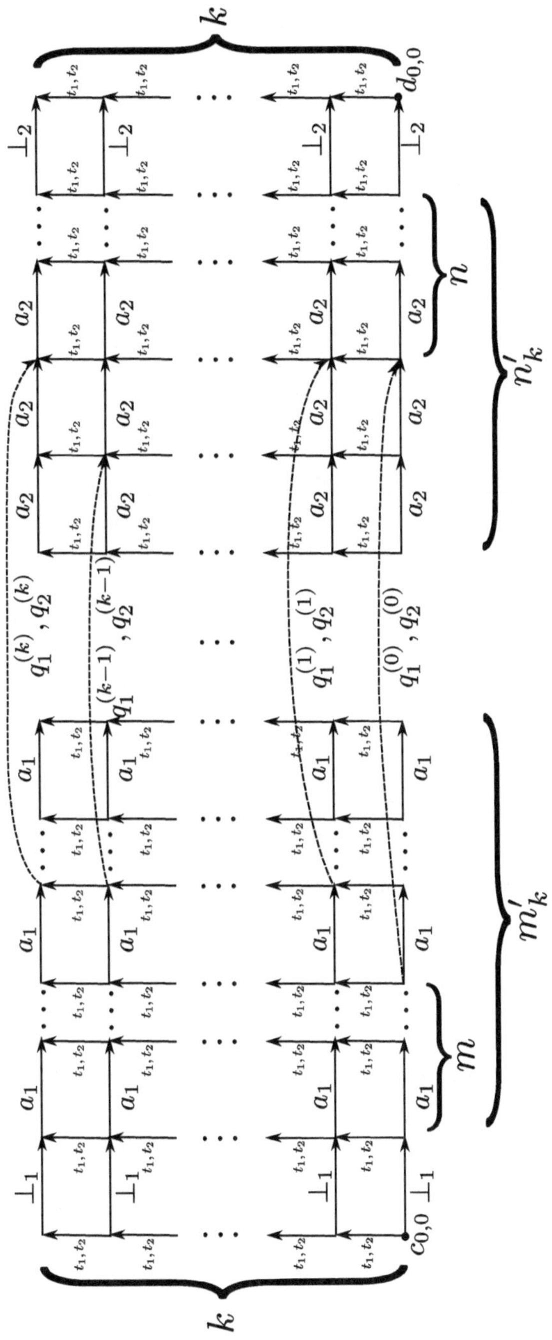

Fig. 3 The automaton $\mathcal{B}^{(k)}_{m,n}$

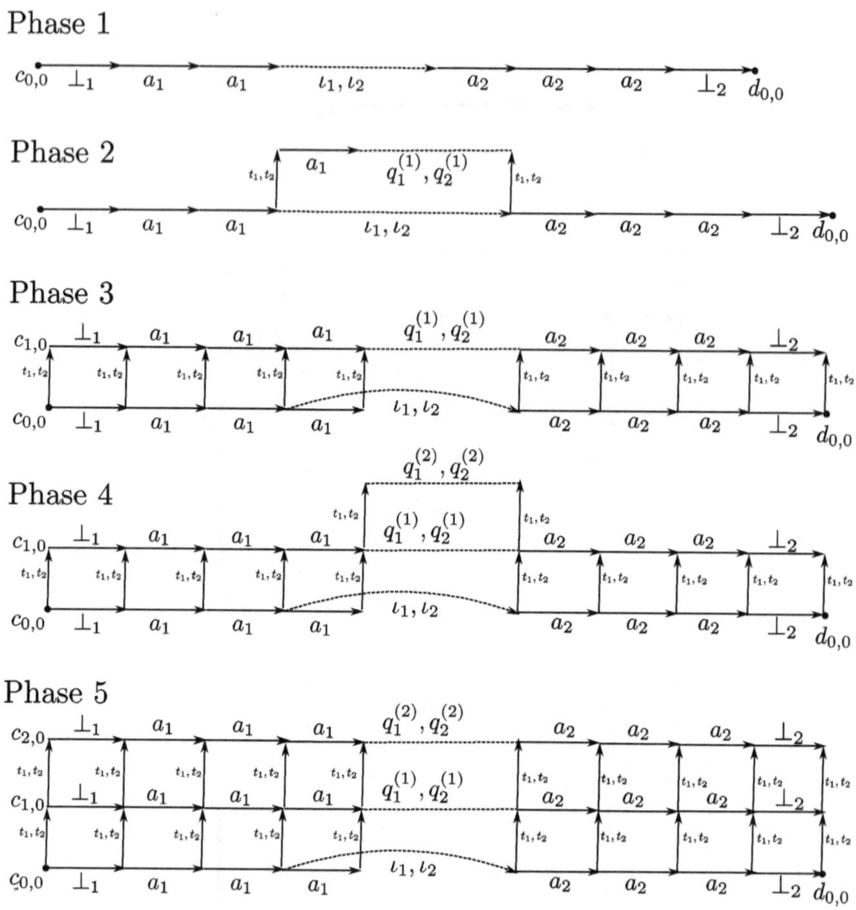

Fig. 4 An example of simulation of the *CM*(2) using Stephen's iterative procedure

$$s\iota = st_1a_1q_1^{(1)}t_1^{-1}, s \in \{a_1, \bot_1\}$$

Thus, we can perform an expansion, which followed by a determination, produces the approximate automaton depicted in Phase 2 of Fig. 4. Now we need to copy the contents of the counters from the first level to the new one. This is achieved via the relations $\mathcal{T}_1^c, \mathcal{T}_2^c$. Indeed, $\mathcal{T}_i^c$ consists of all the relations of the form $t_ix = xt_i$, for $x \in \{a_i, a_i^{-1}, \bot_i, \bot_i^{-1}\}$, for $i = 1, 2$, which applied produces the approximate automaton in Phase 3 of Fig. 4. The second instruction $(q^{(1)}, 2, a, q^{(2)})$ is related to the second counter and the corresponding relations are contained in the set $\mathcal{T}_2^t$ which are:

$$q^{(1)}a_2 = t_2q^{(2)}t_2^{-1}a_2$$

In this case, the test is positive, otherwise if the second counter was empty we would have applied an instruction $(q^{(1)}, 2, \perp_2, p)$ with the corresponding relation $q^{(1)} \perp_2 = t_2 p t_2^{-1} \perp_2$. Performing the expansion and a determination on the second level relative to the relation $q^{(1)} a_2 = t_2 q^{(2)} t_2^{-1} a_2$ produces the approximate automaton depicted in Phase 4 of Fig. 4. Like before, also in this case we need to transfer the content on the counters in the new created level. This is done by performing the expansions and the corresponding determinations of the commutating relations of $\mathcal{T}_1^c, \mathcal{T}_2^c$ (Phase 5 of Fig. 4). In general, it is possible to prove the following lemma.

Lemma 1 *Let $\mathcal{M}$ be a normalized CM(2), let $m, n, k \in \mathbb{N}$, and put $w_{m,n} = \perp_1 a_1^m \iota a_2^n \perp_2$. Then $\mathcal{B}_{m,n}^{(k)}$ is a finite approximate automaton of $\mathcal{A}(X, \mathcal{T}; w_{m,n})$.*

Let $\mathcal{C}$ denote the finite complete inverse automaton with a single vertex and all the loops labeled by all the elements in $X_1 \cup X_2$, (the *bouquet automaton*). The relations contained in $\mathcal{T}_1^f, \mathcal{T}_2^f, \mathcal{T}_U$ force the final state $f = f_1 = f_2$ to be a zero of $S_1 *_U S_2$. Therefore, in case the machine reaches the halting state f at the kth step, the fact that f is zero forces $\mathcal{B}_{m,n}^{(k)}$ to collapse to the bouquet automata $\mathcal{C}$. It follows from the definition that $\mathcal{B}_{m,n}^{(k-1)}$ embeds in $\mathcal{B}_{m,n}^{(k)}$ for every $k \geq 1$, hence we can define $\mathcal{B}_{m,n}$ as the colimit of the sequence $(\mathcal{B}_{m,n}^{(k)})_k$, where all the $\mathcal{B}_{m,n}^{(k)}$ embed. By the previous remark on the bouquet automaton, this colimit may be finite or infinite, depending on whether or not the computation in $\mathcal{M}$ halts when we start with the configuration (ι, m, n). Using the fact that $CM(2)$ is normalized, especially the determinism and the reversibility that forbid that more than one expansion is performed for each edge labeled by some state of Q, and by Lemma 1, it is possible to prove the following proposition.

Proposition 4 *Let $\mathcal{M}$ be a normalized CM(2) and let $m, n \in \mathbb{N}$. Then*

$$\mathcal{A}(X, \mathcal{T}; w_{m,n}) = \begin{cases} \mathcal{C} & \textit{if (m,n) is accepted by } \mathcal{M} \\ \mathcal{B}_{m,n} & \textit{otherwise} \end{cases}$$

Since $\mathcal{C}$ is the Schützenberger automaton of the zero of S, we immediately obtain:

Theorem 6 *Let $\mathcal{M}$ be a normalized CM(2) and let $m, n \in \mathbb{N}$. Then $w_{m,n} = 0$ in the inverse semigroup defined by the associated amalgam $[S_1, S_2; U, \omega_1, \omega_2]$ if and only if (m, n) is accepted by $\mathcal{M}$.*

Hence, if the word problem would be decidable in these circumstances, then, in view of Propositions 2 and 3 and Theorem 6, we could decide whether or not a normalized $CM(2)$ accepts a given $(m, n) \in \mathbb{N}^2$. And the latter is undecidable, even when we consider the single $CM(2)$ $\mathcal{M}^*$ of Corollary 1. Furthermore, if there would be an algorithm to decide whether a $\mathcal{D}$-class of some word is finite or not, then by Proposition 4 it would be possible to decide the halting problem for the machine $\mathcal{M}^*$, and this concludes the sketch of the proof of the two statements (i), (ii) of Theorem 4.

6 Conclusions: The Limit of Decidability of the Word Problem

It is quite clear that the two decidability results illustrated in Sect. 4, can be probably extended to wider classes of amalgams, making use of Theorem 1. However, it seems likely that such classes are not neatly defined, and that there is quite small room between decidability and undecidability results. One could try to deal with decidability with different approaches, for instance in [4] sufficient conditions for amalgamated free products of amalgams satisfying condition 1 of Definition 1 are obtained by a procedure that differs from the one proposed by Bennet mainly in Construction 1. In [10], it is proven that the languages recognized by Schützenberger automata of amalgamated free products of inverse semigroups are context-free, hence the decidability of word problem for such inverse semigroups immediately follows from the decidability of the membership problem for context-free languages. The attempt at classifying languages recognized by Schützenberger automata could be promising, because one could then apply decidability and computational complexity results in the realm of the theory of formal languages. It is however fair to say that to classify these languages in general, one has to use some good information on the shape of the Schützenberger automata.

We would also remark that the shape of Schützenberger automata can also give quite often important information on structural properties of the amalgamated free product, see for instance [3, 6, 26].

Acknowledgments The first author is grateful to the organizers of ICSAOT 2014 for inviting her to present these results and acknowledges support from PRIN project 2011 "Automi e Linguaggi Formali: Aspetti Matematici e Applicativi." The last author acknowledges support from the European Regional Development Fund through the programme COMPETE and by the Portuguese Government through the FCT—Fundação para a Ciência e a Tecnologia under the project PEst-C/MAT/UI0144/2013 as well as support from the FCT project SFRH/BPD/65428/2009.

References

1. Bennett, C.: Logical reversibility of computation. IBM J. Res. Dev. **17**, 525–532 (1973)
2. Bennett, P.: Amalgamated free product of inverse semigroups. J. Algebra **198**, 499–537 (1997)
3. Bennett, P.: On the structure of inverse semigroup amalgams. Int. J. Algebra Comput. **7**(5), 577–604 (1997)
4. Cherubini, A., Mazzucchelli, M.: On the decidability of the word problem for amalgamated free products of inverse semigroups. Semigroup Forum **76**(2), 309–329 (2008)
5. Cherubini, A., Rodaro, E.: Amalgams vs Yamamura's HNN-extensions of inverse semigroups. Algebra Colloq. **18**(04), 647–657 (2011)
6. Cherubini, A., Jajcayová, T., Rodaro, E.: Maximal subgroups of amalgams of finite inverse semigroups. Semigroup Forum **9**(2), 401–424 (2015)
7. Cherubini, A., Meakin, J., Piochi, B.: Amalgams of free inverse semigroups. Semigroup Forum **54**, 199–220 (1997)
8. Cherubini, A., Meakin, J., Piochi, B.: Amalgams of finite inverse semigroups. J. Algebra **285**, 706–725 (2005)

9. Cherubini, A., Nuccio, C., Rodaro, E.: Multilinear equations in amalgams of finite inverse semigroups. Int. J. Algebra Comput. **21**(01n02), 35–59 (2011)
10. Cherubini, A., Nuccio, C., Rodaro, E.: Amalgams of finite inverse semigroups and deterministic context-free languages. Semigroup Forum **85**(1), 129–146 (2012)
11. Haataja, S., Margolis, S., Meakin, J.: Bass-Serre theory for groupoids and the structure of full regular semigroup amalgams. J. Algebra **183**, 38–54 (1996)
12. Hall, T.E.: Finite inverse semigroups and amalgamation. In: Goberstein, S.M., Higgins, P.M. (eds.) Semigroups and Their Applications, pp. 51–56. Reidel, Dordrecht (1987)
13. Jajcayová, T.: HNN-extensions of inverse semigroups. Ph.D. thesis at University of Nebraska-Lincoln Department of Mathematics and Statistics (1997)
14. Janin, D.: Toward a higher-dimensional string string theory for the modeling of computerized systems. Technical report RR-1477-13, LaBRI, IPB, Université de Bordeaux (2013)
15. Jones, P.R., Margolis, S.W., Meakin, J.C., Stephen, J.B.: Free products of inverse semigroups. Glasg. Math. J. **33**, 373–387 (1991)
16. Kellendonk, J.: The local structure of tiling and their integer group of coinvariance. Commun. Math. Phys. **187**, 115–157 (1997)
17. Kellendonk, J.: Topological equivalence of tilings. J. Math. Phys. **38**, 1823–1842 (1997)
18. Kimura, N.: On semigroups. Ph.D. thesis at Tulane University of Louisiana (1957)
19. Lawson, M.V.: Inverse Semigroups. The Theory of Partial Symmetries. World Scientific, River Edge (1998)
20. Margolis, S., Meakin, J., Sapir, M.: Algorithmic problems in groups, semigroups and inverse semigroups. In: Fountain, J. (ed.) Semigroups, Formal Languages and Groups, pp. 147–214 (1995)
21. Meakin, J.: Inverse semigroups: some open questions (2012)
22. Morita, K.: Universality of a reversible two-counter machine. Theor. Comput. Sci. **168**, 303–320 (1996)
23. Novikov, P.S.: On the algorithmic unsolvability of the word problem in group theory. Trudy Matematicheskogo Instituta imeni VA Steklova **3**(29), 44–143 (1955)
24. Paterson, A.L.T.: Grupoids, Inverse Semigroups, and Their Operator Algebras. Birkhauser Boston, Boston (1999)
25. Petrich, M.: Inverse Semigroups. Wiley, New York (1984)
26. Rodaro, E.: Bicyclic subsemigroups in amalgams of finite inverse semigroups. Int. J. Algebra Comput. **20**(1), 89–113 (2010)
27. Rodaro, E., Cherubini, A.: Decidability of the word problem in Yamamura's HNN-extensions of finite inverse semigroups. Semigroup Forum **77**(2), 163–186 (2008)
28. Rodaro, E., Silva, P.V.: Amalgams of inverse semigroups and reversible two-counter machines. J. Pure Appl. Algebra **217**(4), 585–597 (2013)
29. Sapir, M.V.: Algorithmic problems for amalgams of finite semigroups. J. Algebra **229**(2), 514–531 (2000)
30. Steinberg, B.: A topological approach to inverse and regular semigroups. Pac. J. Math. **208**(2), 367–396 (2003)
31. Stephen, J.B.: Presentation of inverse monoids. J. Pure Appl. Algebra **198**, 81–112 (1990)
32. Stephen, J.B.: Amalgamated free products of inverse semigroups. J. Algebra **208**, 339–424 (1998)

A Nonfinitely Based Semigroup of Triangular Matrices

M.V. Volkov

Abstract A new sufficient condition under which a semigroup admits no finite identity basis has been recently suggested in a joint paper by Karl Auinger, Yuzhu Chen, Xun Hu, Yanfeng Luo, and the author. Here we apply this condition to show the absence of a finite identity basis for the semigroup $\mathrm{UT}_3(\mathbb{R})$ of all upper triangular real 3×3-matrices with 0 s and/or 1 s on the main diagonal. The result holds also for the case when $\mathrm{UT}_3(\mathbb{R})$ is considered as an involution semigroup under the reflection with respect to the secondary diagonal.

Keywords Semigroup reduct · Involution semigroup · semigroup variety

1 Introduction

A *semigroup identity* is just a pair of *words*, i.e., elements of the free semigroup A^+ over an alphabet A. In this paper, identities are written as "bumped" equalities such as $u \bumpeq v$. The identity $u \bumpeq v$ is *trivial* if $u = v$ and *nontrivial* otherwise. A semigroup S *satisfies* $u \bumpeq v$ where $u, v \in A^+$ if for every homomorphism $\varphi : A^+ \to S$, the equality $u\varphi = v\varphi$ is valid in S; alternatively, we say that $u \bumpeq v$ *holds* in S. Clearly, trivial identities hold in every semigroup, and there exist semigroups (for instance, free semigroups over non-singleton alphabets) that satisfy only trivial identities.

Given any system Σ of semigroup identities, we say that an identity $u \bumpeq v$ *follows* from Σ if every semigroup satisfying all identities of Σ satisfies the identity $u \bumpeq v$ as well; alternatively, we say that Σ *implies* $u \bumpeq v$. A semigroup S is said to be *finitely based* if there exists a finite identity system Σ such that every identity holding in S follows from Σ; otherwise, S is called *nonfinitely based*.

The finite basis problem, that is, the problem of classifying semigroups according to the finite basability of their identities, has been intensively explored since the mid-

M.V. Volkov (✉)
Institute of Mathematics and Computer Science, Ural Federal University,
Lenina 51, 620000 Ekaterinburg, Russia
e-mail: mikhail.volkov@usu.ru

P.G. Romeo et al. (eds.), *Semigroups, Algebras and Operator Theory*,
Springer Proceedings in Mathematics & Statistics 142,
DOI 10.1007/978-81-322-2488-4_2

1960s when the very first examples of nonfinitely based semigroups were discovered by Austin [3], Biryukov [5], and Perkins [15, 16]. One of the examples by Perkins was especially impressive as it involved a very transparent and natural object. Namely, Perkins proved that the finite basis property fails for the 6-element semigroup formed by the following integer 2×2-matrices under the usual matrix multiplication:

$$\begin{pmatrix}0&0\\0&0\end{pmatrix}, \begin{pmatrix}1&0\\0&0\end{pmatrix}, \begin{pmatrix}0&1\\0&0\end{pmatrix}, \begin{pmatrix}0&0\\1&0\end{pmatrix}, \begin{pmatrix}0&0\\0&1\end{pmatrix}, \begin{pmatrix}1&0\\0&1\end{pmatrix}.$$

Thus, even a finite semigroup can be nonfinitely based; moreover, it turns out that semigroups are the only "classical" algebras for which finite nonfinitely based objects can exist: finite groups [14], finite associative and Lie rings [4, 10, 11], and finite lattices [13] are all finitely based. Therefore studying finite semigroups from the viewpoint of the finite basis problem has become a hot area in which many neat results have been achieved and several powerful methods have been developed, see the survey [18] for an overview.

It may appear surprising but the finite basis problem for **infinite** semigroups is less studied. The reason for this is that infinite semigroups usually arise in mathematics as semigroups of transformations of an infinite set, or semigroups of relations on an infinite domain, or semigroups of matrices over an infinite ring, and as a rule all these semigroups are "too big" to satisfy any nontrivial identity. For instance (see, e.g., [6]), the two integer matrices

$$\begin{pmatrix}2&0\\&1\end{pmatrix}, \begin{pmatrix}2&1\\&1\end{pmatrix}$$

are known to generate a free subsemigroup in the semigroup $T_2(\mathbb{Z})$ of all upper triangular integer 2×2-matrices. (Here and below we omit zero entries under the main diagonal when dealing with upper triangular matrices.) Therefore even such a "small" matrix semigroup as $T_2(\mathbb{Z})$ satisfies only trivial identities, to say nothing about matrix semigroups of larger dimension.

If all identities holding in a semigroup S are trivial, S is finitely based in a void way, so to speak. If, however, an infinite semigroup satisfies a nontrivial identity, its finite basis problem may constitute a challenge since "finite" methods are nonapplicable in general. Therefore, up to recently, results classifying finitely based and nonfinitely based members within natural families of concrete infinite semigroups that contain semigroups with a nontrivial identity have been rather sparse.

Auinger et al. [1] have found a new sufficient condition under which a semigroup is nonfinitely based and applied this condition to certain important classes of infinite semigroups. In the present paper, we demonstrate yet another application; its interesting feature is that it requires the full strength of the main result of [1]. Namely, we prove that the semigroup $UT_3(\mathbb{R})$ of all upper triangular real 3×3-matrices whose main diagonal entries are 0 s and/or 1 s is nonfinitely based. The result holds also for the case when $UT_3(\mathbb{R})$ is considered as an involution semigroup under the reflection with respect to the secondary diagonal.

The paper is structured as follows. In Sect. 2, we recall the main result from [1], and in Sect. 3 we apply it to the semigroup $UT_3(\mathbb{R})$. Section 4 collects some concluding remarks and a related open question.

An effort has been made to keep this paper self-contained, to a reasonable extent. We use only the most basic concepts of semigroup theory and universal algebra that all can be found in the early chapters of the textbooks [7, 8], a suitable version of the main theorem from [1], and a few minor results from [2, 12, 17].

2 A Sufficient Condition for the Nonexistence of a Finite Basis

The sufficient condition for the nonexistence of a finite basis established in [1] applies to both plain semigroups, i.e., semigroups treated as algebras of type (2), and semigroups with involution as algebras of type (2,1). Let us recall all the concepts needed to formulate this condition.

We start with the definition of an involution semigroup. An algebra $\langle S, \cdot, ^\star\rangle$ of type (2,1) is called an *involution semigroup* if $\langle S, \cdot\rangle$ is a semigroup (referred to as the *semigroup reduct* of $\langle S, \cdot, ^\star\rangle$) and the unary operation $x \mapsto x^\star$ is an involutory anti-automorphism of $\langle S, \cdot\rangle$, that is,

$$(xy)^\star = y^\star x^\star \text{ and } (x^\star)^\star = x$$

for all $x, y \in S$.

The *free involution semigroup* $\mathcal{FI}(A)$ on a given alphabet A can be constructed as follows. Let $\overline{A} := \{a^\star \mid a \in A\}$ be a disjoint copy of A. Define $(a^\star)^\star := a$ for all $a^\star \in \overline{A}$. Then $\mathcal{FI}(A)$ is the free semigroup $(A \cup \overline{A})^+$ endowed with the involution defined by

$$(a_1 \cdots a_m)^\star := a_m^\star \cdots a_1^\star$$

for all $a_1, \ldots, a_m \in A \cup \overline{A}$. We refer to elements of $\mathcal{FI}(A)$ as *involutory words over* A. An *involutory identity* $u \simeq v$ is just a pair of involutory words; the identity *holds* in an involution semigroup S if for every involution semigroup homomorphism $\varphi : \mathcal{FI}(A)^+ \to S$, the equality $u\varphi = v\varphi$ is valid in S. Now the concepts of a finitely based/nonfinitely based involution semigroup are defined exactly as in the plain semigroup case. In what follows, we use square brackets to indicate adjustments to be made in the involution case.

A class $\mathbf{V}$ of [involution] semigroups is called a *variety* if there exits a system Σ of [involution] semigroup identities such that $\mathbf{V}$ consists precisely of all [involution] semigroups that satisfy every identity in Σ. In this case we say that $\mathbf{V}$ is *defined* by Σ. If the system Σ can be chosen to be finite, the corresponding variety is said to be *finitely based*; otherwise it is *nonfinitely based*. Given a class $\mathbf{K}$ of [involution] semigroups, the variety defined by the identities that hold in each [involution] semigroup from $\mathbf{K}$ is said to be *generated by* $\mathbf{K}$ and is denoted by $\mathsf{var}\,\mathbf{K}$; if $\mathbf{K} = \{S\}$, we write $\mathsf{var}\,S$ rather than $\mathsf{var}\{S\}$. It should be clear that S and $\mathsf{var}\,S$ are simultaneously finitely based or nonfinitely based.

A semigroup is said to be *periodic* if each of its one-generated subsemigroups is finite and *locally finite* if each of its finitely generated subsemigroups is finite. A variety of semigroups is *locally finite* if all its members are locally finite.

Let $\mathbf{A}$ and $\mathbf{B}$ be two classes of semigroups. The *Mal'cev product* $\mathbf{A} \circledm \mathbf{B}$ of $\mathbf{A}$ and $\mathbf{B}$ is the class of all semigroups S for which there exists a congruence θ such that the quotient semigroup S/θ lies in $\mathbf{B}$ while all θ-classes that are subsemigroups in S belong to $\mathbf{A}$. Notice that for a congruence θ on a semigroup S, a θ-class forms a subsemigroup of S if and only if the class is an idempotent of the quotient semigroup S/θ.

Let $x_1, x_2, \ldots, x_n, \ldots$ be a sequence of letters. The sequence $\{Z_n\}_{n=1,2,\ldots}$ of *Zimin words* is defined inductively by $Z_1 := x_1$, $Z_{n+1} := Z_n x_{n+1} Z_n$. We say that a word v is an [*involutory*] *isoterm* for a class $\mathbf{C}$ of semigroups [with involution] if the only [involutory] word v' such that all members of $\mathbf{C}$ satisfy the [involution] semigroup identity $v \bumpeq v'$ is the word v itself.

Now we are in a position to state the main result of [1]. Here **Com** denotes the variety of all commutative semigroups.

Theorem 1 ([1, Theorem 6]) *A variety* $\mathbf{V}$ *of* [*involution*] *semigroups is nonfinitely based provided that*

(i) [*the class of all semigroup reducts of*] $\mathbf{V}$ *is contained in the variety* $\mathsf{var}(\mathbf{Com} \circledm \mathbf{W})$ *for some locally finite semigroup variety* $\mathbf{W}$ *and*
(ii) *each Zimin word is an* [*involutory*] *isoterm relative to* $\mathbf{V}$.

Formulated as above, Theorem 1 suffices for all applications presented in [1] but is insufficient for the purposes of the present paper. However, it is observed in [1, Remark 1] that the theorem remains valid if one replaces the condition (i) by the following weaker condition:

(i′) [the class of all semigroup reducts of] $\mathbf{V}$ is contained in the variety $\mathsf{var}(\mathbf{U} \circledm \mathbf{W})$ where $\mathbf{U}$ is a semigroup variety all of whose periodic members are locally finite and $\mathbf{W}$ is a locally finite semigroup variety.

Here we will utilize this stronger form of Theorem 1.

3 The Identities of $\mathrm{UT}_3(\mathbb{R})$

Recall that we denote by $\mathrm{UT}_3(\mathbb{R})$ the semigroup of all upper triangular real 3×3-matrices whose main diagonal entries are 0 s and/or 1 s. For each matrix $\alpha \in \mathrm{UT}_3(\mathbb{R})$, let α^D stand for the matrix obtained by reflecting α with respect to the secondary diagonal (from the top right to the bottom left corner); in other words, $(\alpha_{ij})^D := (\alpha_{4-j\,4-i})$. Then it is easy to verify that the unary operation $\alpha \mapsto \alpha^D$ (called the *skew transposition*) is an involutory anti-automorphism of $\mathrm{UT}_3(\mathbb{R})$. Thus, we can consider $\mathrm{UT}_3(\mathbb{R})$ also as an involution semigroup. Our main result is the following

Theorem 2 *The semigroup* $\mathrm{UT}_3(\mathbb{R})$ *is nonfinitely based as both a plain semigroup and an involution semigroup under the skew transposition.*

Proof We will verify that the [involution] semigroup variety $\mathsf{var}\,\mathrm{UT}_3(\mathbb{R})$ satisfies the conditions (i′) and (ii) discussed at the end of Sect. 2; the desired result will then follow from Theorem 1 in its stronger form.

Let D_3 denote the 8-element subsemigroup of $\mathrm{UT}_3(\mathbb{R})$ consisting of all diagonal matrices. To every matrix $\alpha \in \mathrm{UT}_3(\mathbb{R})$ we assign the diagonal matrix $\mathsf{Diag}(\alpha) \in \mathrm{D}_3$ by changing each nondiagonal entry of α to 0. The following observation is obvious.

Lemma 3 *The map* $\alpha \mapsto \mathsf{Diag}(\alpha)$ *is a homomorphism of* $\mathrm{UT}_3(\mathbb{R})$ *onto* D_3.

We denote by θ the kernel of the homomorphism of Lemma 3, i.e.,

$$(\alpha, \beta) \in \theta \text{ if and only if } \mathsf{Diag}(\alpha) = \mathsf{Diag}(\beta).$$

Then θ is a congruence on $\mathrm{UT}_3(\mathbb{R})$. Since each element of the semigroup D_3 is an idempotent, each θ-class is a subsemigroup of $\mathrm{UT}_3(\mathbb{R})$. The next fact is the core of our proof.

Proposition 4 *Each* θ*-class of* $\mathrm{UT}_3(\mathbb{R})$ *satisfies the identity*

$$Z_4 \doteq (x_1x_2)^2x_1x_3x_1x_4x_1x_3x_1(x_2x_1)^2. \tag{1}$$

Proof We have to consider eight cases. First we observe that the identity (1) is left-right symmetric, and therefore, (1) holds in some subsemigroup S of $\mathrm{UT}_3(\mathbb{R})$ if and only if it holds in the subsemigroup $S^D = \{s^D \mid s \in S\}$ since S^D is anti-isomorphic to S. This helps us to shorten the below analysis.

Case 1: $S_{000} = \left\{\begin{pmatrix} 0 & \alpha_{12} & \alpha_{13} \\ & 0 & \alpha_{23} \\ & & 0 \end{pmatrix} \mid \alpha_{12}, \alpha_{13}, \alpha_{23} \in \mathbb{R}\right\}$. This subsemigroup is easily seen to satisfy the identity $x_1x_2x_3 \doteq y_1y_2y_3$ which clearly implies (1).

Case 2: $S_{100} = \left\{\begin{pmatrix} 1 & \alpha_{12} & \alpha_{13} \\ & 0 & \alpha_{23} \\ & & 0 \end{pmatrix} \mid \alpha_{12}, \alpha_{13}, \alpha_{23} \in \mathbb{R}\right\}$. Multiplying three arbitrary matrices $\alpha, \beta, \gamma \in S_{100}$, we get

$$\begin{pmatrix} 1 & \alpha_{12} & \alpha_{13} \\ & 0 & \alpha_{23} \\ & & 0 \end{pmatrix} \cdot \begin{pmatrix} 1 & \beta_{12} & \beta_{13} \\ & 0 & \beta_{23} \\ & & 0 \end{pmatrix} \cdot \begin{pmatrix} 1 & \gamma_{12} & \gamma_{13} \\ & 0 & \gamma_{23} \\ & & 0 \end{pmatrix} =$$

$$\begin{pmatrix} 1 & \gamma_{12} & \gamma_{13} + \beta_{12}\gamma_{23} \\ & 0 & 0 \\ & & 0 \end{pmatrix} = \begin{pmatrix} 1 & \beta_{12} & \beta_{13} \\ & 0 & \beta_{23} \\ & & 0 \end{pmatrix} \cdot \begin{pmatrix} 1 & \gamma_{12} & \gamma_{13} \\ & 0 & \gamma_{23} \\ & & 0 \end{pmatrix}.$$

Thus, $\alpha\beta\gamma = \beta\gamma$ and we have proved that S_{100} satisfies the identity $xyz \simeq yz$. Clearly, this identity implies (1).

Case 3: $S_{010} = \left\{\begin{pmatrix} 0 & \alpha_{12} & \alpha_{13} \\ & 1 & \alpha_{23} \\ & & 0 \end{pmatrix} \mid \alpha_{12}, \alpha_{13}, \alpha_{23} \in \mathbb{R}\right\}$. It is easy to see that this subsemigroup satisfies the identity $xyx \simeq x$ which clearly implies (1).

Case 4: $S_{001} = \left\{\begin{pmatrix} 0 & \alpha_{12} & \alpha_{13} \\ & 0 & \alpha_{23} \\ & & 1 \end{pmatrix} \mid \alpha_{12}, \alpha_{13}, \alpha_{23} \in \mathbb{R}\right\}$. This case reduces to Case 2 since $S_{001} = S_{100}^D$.

Case 5: $S_{110} = \left\{\begin{pmatrix} 1 & \alpha_{12} & \alpha_{13} \\ & 1 & \alpha_{23} \\ & & 0 \end{pmatrix} \mid \alpha_{12}, \alpha_{13}, \alpha_{23} \in \mathbb{R}\right\}$. Multiplying three arbitrary matrices $\alpha, \beta, \gamma \in S_{110}$, we get

$$\begin{pmatrix} 1 & \alpha_{12} & \alpha_{13} \\ & 1 & \alpha_{23} \\ & & 0 \end{pmatrix} \cdot \begin{pmatrix} 1 & \beta_{12} & \beta_{13} \\ & 1 & \beta_{23} \\ & & 0 \end{pmatrix} \cdot \begin{pmatrix} 1 & \gamma_{12} & \gamma_{13} \\ & 1 & \gamma_{23} \\ & & 0 \end{pmatrix} = \begin{pmatrix} 1 & \alpha_{12} + \beta_{12} + \gamma_{12} & \gamma_{13} + (\alpha_{12} + \beta_{12})\gamma_{23} \\ & 1 & \gamma_{23} \\ & & 0 \end{pmatrix}$$

whence the product $\alpha\beta\gamma$ depends only on γ and on the sum $\alpha_{12} + \beta_{12}$. Thus, $\alpha\beta\gamma = \beta\alpha\gamma$ and we have proved that S_{110} satisfies the identity $xyz \simeq yxz$. This identity implies (1).

Case 6: $S_{101} = \left\{\begin{pmatrix} 1 & \alpha_{12} & \alpha_{13} \\ & 0 & \alpha_{23} \\ & & 1 \end{pmatrix} \mid \alpha_{12}, \alpha_{13}, \alpha_{23} \in \mathbb{R}\right\}$. Take an arbitrary homomorphism $\varphi : \{x_1, x_2, x_3, x_4\}^+ \to S_{101}$ and let $\alpha = x_1\varphi$, $\beta = x_2\varphi$, $\gamma = x_3\varphi$, and $\delta = x_4\varphi$. Then one can calculate that both $Z_4\varphi$ and $(x_1x_2x_1x_2x_1x_3x_1x_4x_1x_3x_1x_2x_1x_2x_1)\varphi$ are equal to the matrix $\begin{pmatrix} 1 & \alpha_{12} & \varepsilon \\ & 0 & \alpha_{23} \\ & & 1 \end{pmatrix}$ where ε stands for the following expression:

$$8\alpha_{13} + 4\beta_{13} + 2\gamma_{13} + \delta_{13} + \alpha_{12}(4\beta_{23} + 2\gamma_{23} + \delta_{23}) + (4\beta_{12} + 2\gamma_{12} + \delta_{12})\alpha_{23}.$$

Thus, the identity (1) holds on S_{101}.

For readers familiar with the Rees matrix construction (cf. [8, Chap. 3]), we outline a more conceptual proof for the fact that S_{101} satisfies (1). Let $G = \langle \mathbb{R}, + \rangle$ stand for the additive group of real numbers and let P be the $\mathbb{R}\times\mathbb{R}$-matrix over G whose element in the rth row and the sth column is equal to rs. One readily verifies that the map $\begin{pmatrix} 1 & \alpha_{12} & \alpha_{13} \\ & 0 & \alpha_{23} \\ & & 1 \end{pmatrix} \mapsto (\alpha_{23}, \alpha_{13}, \alpha_{12})$ constitutes an isomorphism of the semigroup S_{101} onto the Rees matrix semigroup $\mathrm{M}(\mathbb{R}, G, \mathbb{R}; P)$. It is known (see, e.g., [9]) and easy to verify that every Rees matrix semigroup over an Abelian group satisfies each identity $u \simeq v$ for which the following three conditions hold: the first letter of u is the same as the first letter of v; the last letter of u is the same as the last letter of v; for each ordered pair of letters, the number of occurrences of this pair is the same in u and v. Inspecting the identity (1), one immediately sees that it satisfies the three conditions whence it holds in the semigroup $M(\mathbb{R}, G, \mathbb{R}; P)$ and also in the semigroup S_{101} isomorphic to $M(\mathbb{R}, G, \mathbb{R}; P)$.

Case 7: $S_{011} = \left\{ \begin{pmatrix} 0 & \alpha_{12} & \alpha_{13} \\ & 1 & \alpha_{23} \\ & & 1 \end{pmatrix} \mid \alpha_{12}, \alpha_{13}, \alpha_{23} \in \mathbb{R} \right\}$. This case reduces to Case 5 since $S_{011} = S_{110}^D$.

Case 8: $S_{111} = \left\{ \begin{pmatrix} 1 & \alpha_{12} & \alpha_{13} \\ & 1 & \alpha_{23} \\ & & 1 \end{pmatrix} \mid \alpha_{12}, \alpha_{13}, \alpha_{23} \in \mathbb{R} \right\}$. The semigroup S_{111} is in fact the group of all real upper unitriangular 3×3-matrices. The latter group is known to be nilpotent of class 2, and as observed by Mal'cev [12], every nilpotent group of class 2 satisfies the semigroup identity

$$xzytyzx \doteq yzxtxzy. \tag{2}$$

Now we verify that (1) follows from (2). For this, we substitute in (2) the letter x_1 for x, the letter x_3 for z, the word $x_1x_2x_1$ for y, and the letter x_4 for t. We then obtain the identity

$$\underbrace{x_1}_{x}\,\underbrace{x_3}_{z}\,\underbrace{x_1x_2x_1}_{y}\,\underbrace{x_4}_{t}\,\underbrace{x_1x_2x_1}_{y}\,\underbrace{x_3}_{z}\,\underbrace{x_1}_{x} \doteq \underbrace{x_1x_2x_1}_{y}\,\underbrace{x_3}_{z}\,\underbrace{x_1}_{x}\,\underbrace{x_4}_{t}\,\underbrace{x_1}_{x}\,\underbrace{x_3}_{z}\,\underbrace{x_1x_2x_1}_{y}.$$

Multiplying this identity through by x_1x_2 on the left and by x_2x_1 on the right, we get (1). □

Recall that a semigroup identity $u \doteq v$ is said to be *balanced* if for every letter the number of occurrences of this letter is the same in u and v. Clearly, the identity (1) is balanced.

Lemma 5 ([17, Lemma 3.3]) *If a semigroup variety* $\mathbf{V}$ *satisfies a nontrivial balanced identity of the form* $Z_n \doteq v$, *then all periodic members of* $\mathbf{V}$ *are locally finite.*

Let $\mathbf{U}$ stand for the semigroup variety defined by the identity (1). Then Lemma 5 ensures that all periodic members of $\mathbf{U}$ are locally finite while Lemma 3 and Proposition 4 imply that the semigroup $UT_3(\mathbb{R})$ lies in the Mal'cev product $\mathbf{U} \circledm \mathsf{var}\, D_3$. The variety $\mathsf{var}\, D_3$ is locally finite as a variety generated by a finite semigroup [7, Theorem 10.16]. We see that the variety $\mathsf{var}\, UT_3(\mathbb{R})$ satisfies the condition (i′).

It remains to verify that $\mathsf{var}\, UT_3(\mathbb{R})$ satisfies the condition (ii) as well. Clearly, the involutory version of the condition (ii) is stronger than its plain version so that it suffices to show that each Zimin word is an involutory isoterm relative to $UT_3(\mathbb{R})$ considered as an involution semigroup.

Let TA_2^1 stand for the involution semigroup formed by the (0, 1)-matrices

$$\begin{pmatrix} 0 & 0 \\ 0 & 0 \end{pmatrix}, \begin{pmatrix} 1 & 0 \\ 0 & 0 \end{pmatrix}, \begin{pmatrix} 0 & 1 \\ 0 & 0 \end{pmatrix}, \begin{pmatrix} 1 & 0 \\ 1 & 0 \end{pmatrix}, \begin{pmatrix} 0 & 1 \\ 0 & 1 \end{pmatrix}, \begin{pmatrix} 1 & 0 \\ 0 & 1 \end{pmatrix}$$

under the usual matrix multiplication and the unary operation that swaps each of the matrices $\left(\begin{smallmatrix}1&0\\0&0\end{smallmatrix}\right)$ and $\left(\begin{smallmatrix}0&1\\0&1\end{smallmatrix}\right)$ with the other one and fixes the rest four matrices. By [2, Corollaries 2.7 and 2.8] each Zimin word is an involutory isoterm relative to TA_2^1. Now consider the involution subsemigroup M in $\mathrm{UT}_3(\mathbb{R})$ generated by the matrices

$$e = \begin{pmatrix}1&0&0\\&1&0\\&&1\end{pmatrix}, \quad x = \begin{pmatrix}1&0&0\\&0&0\\&&1\end{pmatrix}, \quad \text{and} \quad y = \begin{pmatrix}1&1&0\\&0&1\\&&1\end{pmatrix}.$$

Clearly, for each matrix $(\mu_{ij}) \in M$, one has $\mu_{ij} \geq 0$ and $\mu_{11} = \mu_{33} = 1$, whence the set N of all matrices $(\mu_{ij}) \in M$ such that $\mu_{13} > 0$ forms an ideal in M. Clearly, N is closed under the skew transposition. A straightforward calculation shows that, besides e, x, and y, the only matrices in $M \backslash N$ are $xy = \left(\begin{smallmatrix}1&1&0\\&0&0\\&&1\end{smallmatrix}\right)$ and $yx = \left(\begin{smallmatrix}1&0&0\\&0&1\\&&1\end{smallmatrix}\right)$. Consider the following bijection between $M \backslash N$ and the set of nonzero matrices in TA_2^1:

$$e \mapsto \left(\begin{smallmatrix}1&0\\0&1\end{smallmatrix}\right), \quad x \mapsto \left(\begin{smallmatrix}1&0\\1&0\end{smallmatrix}\right), \quad y \mapsto \left(\begin{smallmatrix}0&1\\0&0\end{smallmatrix}\right), \quad xy \mapsto \left(\begin{smallmatrix}0&1\\0&1\end{smallmatrix}\right), \quad yx \mapsto \left(\begin{smallmatrix}1&0\\0&0\end{smallmatrix}\right).$$

Extending this bijection to M by sending all elements from N to $\left(\begin{smallmatrix}0&0\\0&0\end{smallmatrix}\right)$ yields an involution semigroup homomorphism from M onto TA_2^1. Thus, TA_2^1 as a homomorphic image of an involution subsemigroup in $\mathrm{UT}_3(\mathbb{R})$ satisfies all involution semigroup identities that hold in $\mathrm{UT}_3(\mathbb{R})$. Therefore, each Zimin word is an involutory isoterm relative to $\mathrm{UT}_3(\mathbb{R})$, as required. □

4 Concluding Remarks and an Open Question

Here we discuss which conditions of Theorem 2 are essential and which can be relaxed.

It should be clear from the above proof of Theorem 2 that the fact that we have dealt with matrices over the field $\mathbb{R}$ is not really essential: the proof works for every semigroup of the form $\mathrm{UT}_3(R)$ where R is an arbitrary associative and commutative ring with 1 such that

$$\underbrace{1 + 1 + \cdots + 1}_{n \text{ times}} \neq 0 \tag{3}$$

for every positive integer n. For instance, we can conclude that the semigroup $\mathrm{UT}_3(\mathbb{Z})$ of all upper triangular integer 3×3-matrices whose main diagonal entries are 0s and/or 1s is nonfinitely based in both plain and involution semigroup settings.

On the other hand, we cannot get rid of the restriction imposed on the main diagonal entries: as the example reproduced in the introduction implies, the semigroup $\mathrm{T}_3(\mathbb{Z})$ of all upper triangular integer 3×3-matrices is finitely based as a plain semigroup since it satisfies only trivial semigroup identities. In a similar way, one can show that $T_3(\mathbb{Z})$ is finitely based when considered as an involution semigroup with the skew transposition. Indeed, the subsemigroup generated in $T_3(\mathbb{Z})$ by the matrix $\zeta = \begin{pmatrix} 2 & 0 & 0 \\ & 1 & 1 \\ & & 2 \end{pmatrix}$ and its skew transpose ζ^D is free, and therefore, considered as an involution semigroup, it is isomorphic to the free involution semigroup on one generator, say z. However, $\mathcal{FI}(\{z\})$ contains as an involution subsemigroup a free involutory semigroup on countably many generators, namely, $\mathcal{FI}(Z)$ where

$$Z = \{zz^*z,\ z(z^*)^2z,\ \dots,\ z(z^*)^nz,\ \dots\}.$$

Hence $T_3(\mathbb{Z})$ satisfies only trivial involution semigroup identities. Of course, the same conclusions persist if we substitute $\mathbb{Z}$ by any associative and commutative ring with 1 satisfying (3) for every n.

We can, however, slightly weaken the restriction on the main diagonal entries by allowing them to take values in the set $\{0, \pm 1\}$. The proof of Theorem 2 remains valid for the resulting [involution] semigroup that we denote by $\mathrm{UT}_3^{\pm}(\mathbb{R})$. Indeed, the homomorphism $\alpha \mapsto \mathsf{Diag}(\alpha)$ of Lemma 3 extends to a homomorphism of $\mathrm{UT}_3^{\pm}(\mathbb{R})$ onto its 27-element subsemigroup consisting of diagonal matrices. The subsemigroup classes of the kernel of this homomorphism are precisely the subsemigroups $S_{000}, \dots, S_{111}$ from the proof of Proposition 4, and therefore, the variety $\mathsf{var}\,\mathrm{UT}_3^{\pm}(\mathbb{R})$ satisfies the condition (i′) of the stronger form of Theorem 1. Of course, the variety fulfills also the condition (ii) since (ii) is inherited by supervarieties. In the same fashion, the proof of Theorem 2 applies, say, to the semigroup of all upper triangular complex 3×3-matrices whose main diagonal entries come from the set $\{0, 1, \xi, \dots, \xi^{n-1}\}$ where ξ is a primitive nth root of unity.

The question of whether or not a result similar to Theorem 2 holds true for analogs of the semigroup $\mathrm{UT}_3(\mathbb{R})$ in other dimensions is more involved. The variety $\mathsf{var}\,\mathrm{UT}_2(\mathbb{R})$ fulfills the condition (i′) since the condition is clearly inherited by subvarieties and the injective map $\mathrm{UT}_2(\mathbb{R}) \to \mathrm{UT}_3(\mathbb{R})$ defined by $\begin{pmatrix} \alpha_{11} & \alpha_{12} \\ & \alpha_{22} \end{pmatrix} \mapsto \begin{pmatrix} \alpha_{11} & 0 & \alpha_{12} \\ & 0 & 0 \\ & & \alpha_{22} \end{pmatrix}$ is an embedding of [involution] semigroups whence $\mathsf{var}\,\mathrm{UT}_2(\mathbb{R}) \subseteq \mathsf{var}\,\mathrm{UT}_3(\mathbb{R})$. However, $\mathsf{var}\,\mathrm{UT}_2(\mathbb{R})$ does not satisfy the condition (ii) as the following result shows.

Proposition 6 *The semigroup* $\mathrm{UT}_2(\mathbb{R})$ *of all upper triangular real* 2×2*-matrices whose main diagonal entries are* 0*s and/or* 1*s satisfies the identity*

$$Z_4 \simeq x_1x_2x_1x_3x_1^2x_2x_4x_2x_1^2x_3x_1x_2x_1. \tag{4}$$

Proof Fix an arbitrary homomorphism $\varphi : \{x_1, x_2, x_3, x_4\}^+ \to \mathrm{UT}_2(\mathbb{R})$. For brevity, denote the right-hand side of (4) by w; we thus have to prove that $Z_4\varphi = w\varphi$. Let

$$x_1\varphi = \begin{pmatrix} \alpha_{11} & \alpha_{12} \\ & \alpha_{22} \end{pmatrix},\quad x_2\varphi = \begin{pmatrix} \beta_{11} & \beta_{12} \\ & \beta_{22} \end{pmatrix},\quad x_3\varphi = \begin{pmatrix} \gamma_{11} & \gamma_{12} \\ & \gamma_{22} \end{pmatrix},\quad x_4\varphi = \begin{pmatrix} \delta_{11} & \delta_{12} \\ & \delta_{22} \end{pmatrix},$$

where $\alpha_{11}, \alpha_{22}, \beta_{11}, \beta_{22}, \gamma_{11}, \gamma_{22}, \delta_{11}, \delta_{22} \in \{0, 1\}$ and $\alpha_{12}, \beta_{12}, \gamma_{12}, \delta_{12} \in \mathbb{R}$.

If $\alpha_{22} = 0$, the fact that $\alpha_{11}, \beta_{11}, \gamma_{11}, \delta_{11} \in \{0, 1\}$ readily implies that $Z_4\varphi = w\varphi = \left(\begin{smallmatrix} \varepsilon & \varepsilon\alpha_{12} \\ & 0 \end{smallmatrix}\right)$, where $\varepsilon = \alpha_{11}\beta_{11}\gamma_{11}\delta_{11}$. Similarly, if $\alpha_{11} = 0$, then it is easy to calculate that $Z_4\varphi = w\varphi = \left(\begin{smallmatrix} 0 & \alpha_{12}\eta \\ & \eta \end{smallmatrix}\right)$, where $\eta = \alpha_{22}\beta_{22}\gamma_{22}\delta_{22}$. We thus may (and will) assume that $\alpha_{11} = \alpha_{22} = 1$.

Now, if $\beta_{22} = 0$, a straightforward calculation shows that $Z_4\varphi = w\varphi = \left(\begin{smallmatrix} \varkappa & \varkappa(\alpha_{12}+\beta_{12}) \\ & 0 \end{smallmatrix}\right)$, where $\varkappa = \beta_{11}\gamma_{11}\delta_{11}$. Similarly, if $\beta_{11} = 0$, we get $Z_4\varphi = w\varphi = \left(\begin{smallmatrix} 0 & (\alpha_{12}+\beta_{12})\lambda \\ & \lambda \end{smallmatrix}\right)$, where $\lambda = \beta_{22}\gamma_{22}\delta_{22}$. Thus, we may also assume that $\beta_{11} = \beta_{22} = 1$. Observe that the word w is obtained from the word Z_4 by substituting $x_1^2x_2$ for the second occurrence of the factor $x_1x_2x_1$ and $x_2x_1^2$ for the third occurrence of this factor. Therefore $\alpha_{11} = \alpha_{22} = \beta_{11} = \beta_{22} = 1$ implies $x_1x_2x_1\varphi = x_1^2x_2\varphi = x_2x_1^2\varphi = \left(\begin{smallmatrix} 1 & 2\alpha_{12}+\beta_{12} \\ & 1 \end{smallmatrix}\right)$ whence $Z_4\varphi = w\varphi$. □

Now let $\mathrm{UT}_n(\mathbb{R})$ stand for the semigroup of all upper triangular real $n \times n$-matrices whose main diagonal entries are 0s and/or 1s and assume that $n \geq 4$. Here the behavior of the [involution] semigroup variety generated by $\mathrm{UT}_n(\mathbb{R})$ with respect to the conditions of Theorem 1 is in a sense opposite. Namely, it is not hard to show (by using an argument similar to the one utilized in the proof of Theorem 2) that the variety $\mathsf{var}\,\mathrm{UT}_n(\mathbb{R})$ with $n \geq 4$ satisfies the condition (ii). On the other hand, the approach used in the proof of Theorem 2 fails to justify that this variety fulfills (i′). In order to explain this claim, suppose for simplicity that $n = 4$. Then the homomorphism $\alpha \mapsto \mathsf{Diag}(\alpha)$ maps $\mathrm{UT}_4(\mathbb{R})$ onto its 16-element subsemigroup consisting of diagonal matrices which all are idempotent. This induces a partition of $\mathrm{UT}_4(\mathbb{R})$ into 16 subsemigroups, and to mimic the proof of Theorem 2 one should show that all these subsemigroups belong to a variety whose periodic members are locally finite. One of these 16 subsemigroups is nothing but the group of all real upper unitriangular 4×4-matrices. The latter group is known to be nilpotent of class 3, and one might hope to use the identity

$$xzytyzxsyzxtxzy \simeq yzxtxzysxzytyzx, \tag{5}$$

proved by Mal'cev [12] to hold in every nilpotent group of class 3, along the lines of the proof of Proposition 4 where we have invoked Mal'cev's identity holding in each nilpotent group of class 2. However, it is known [19, Theorem 2] that the variety defined by (5) contains infinite finitely generated periodic semigroups. Even though this fact does not yet mean that the condition (i′) fails in $\mathsf{var}\,\mathrm{UT}_4(\mathbb{R})$, it demonstrates that the techniques presented in this paper are not powerful enough to verify whether or not the variety obeys this condition. It seems that this verification constitutes a very difficult task as it is closely connected with Sapir's longstanding conjecture that for each nilpotent group G, periodic members of the semigroup variety $\mathsf{var}\,G$ are locally finite, see [17, Sect. 5].

Back to our discussion, we see that Theorem 1 cannot be applied to the semigroup $\mathrm{UT}_2(\mathbb{R})$ and we are not in a position to apply it to the semigroups $\mathrm{UT}_n(\mathbb{R})$ with $n \geq 4$. Of course, this does not indicate that these semigroups are finitely based—recall that

Theorem 1 is only a sufficient condition for being nonfinitely based. Presently, we do not know which of the semigroups $\mathrm{UT}_n(\mathbb{R})$ with $n \neq 3$ possess the finite basis property, and we conclude the paper with explicitly stating this open question in the anticipation that, over time, looking for an answer might stimulate creating new approaches to the finite basis problem for infinite [involution] semigroups:

Question For which $n \neq 3$ is the [involution] semigroup $\mathrm{UT}_n(\mathbb{R})$ finitely based?

Acknowledgments The author gratefully acknowledges support from the Presidential Programme "Leading Scientific Schools of the Russian Federation", project no. 5161.2014.1, and from the Russian Foundation for Basic Research, project no. 14-01-00524. In particular, the latter project was used to support the author's participation in the International Conference on Semigroups, Algebras and Operator Theory (ICSAOT-2014) held at the Cochin University of Science and Technology as well as his participation in the International Conference on Algebra and Discrete Mathematics (ICADM-2014) held at Government College, Kattappana. Also the author expresses cordial thanks to the organizers of these two conferences, especially to Dr. P. G. Romeo, Convenor of ICSAOT-2014, and Sri. G. N. Prakash, Convenor of ICADM-2014, for their warm hospitality.

References

1. Auinger, K., Chen, Y., Hu, X., Luo, Y., Volkov, M.V.: The finite basis problem for Kauffman monoids. Algebra Universalis, accepted. [A preprint is available under http://arxiv.org/abs/1405.0783.]
2. Auinger, K., Dolinka, I., Volkov, M.V.: Matrix identities involving multiplication and transposition. J. Eur. Math. Soc. **14**, 937–969 (2012)
3. Austin, A.K.: A closed set of laws which is not generated by a finite set of laws. Q. J. Math. Oxf. Ser. (2) **17**, 11–13 (1966)
4. Bahturin, Yu.A., Ol'shanskii, A.Yu.: Identical relations in finite Lie rings. Mat. Sb., N. Ser. **96**(138), 543–559 (1975). [Russian (English trans.) Math. USSR-Sbornik **25**, 507–523 1975)]
5. Biryukov, A.P.: On infinite collections of identities in semigroups. Algebra i Logika **4**(2), 31–32 (1965). [Russian]
6. Blondel, V.D., Cassaigne, J., Karhumäki, J.: Freeness of multiplicative matrix semigroups, problem 10.3. In: Blonde, V.D., Megretski, A. (eds.) Unsolved Problems in Mathematical Systems and Control Theory, pp. 309–314. Princeton University Press, Princeton (2004)
7. Burris, S., Sankappanavar, H.P.: A Course in Universal Algebra. Springer, New York (1981)
8. Clifford, A.H., Preston, G.B.: The Algebraic Theory of Semigroups, vol. 1. American Mathematical Society, Providence (1961)
9. Kim, K.H., Roush, F.: The semigroup of adjacency patterns of words. Algebraic Theory of Semigroups. Colloqium Mathematical Society János Bolyai, vol. 20, pp. 281–297. North-Holland, Amsterdam (1979)
10. Kruse, R.L.: Identities satisfied by a finite ring. J. Algebra **26**, 298–318 (1973)
11. L'vov, I.V.: Varieties of associative rings. I. Algebra i Logika **12**, 269–297 (1973). [Russian (English trans.) Algebra and Logic **12**, 150–167 (1973)]
12. Mal'cev, A.I.: Nilpotent semigroups. Ivanov. Gos. Ped. Inst. Uč. Zap. Fiz.-Mat. Nauki **4**, 107–111 (1953). [Russian]
13. McKenzie, R.N.: Equational bases for lattice theories. Math. Scand. **27**, 24–38 (1970)
14. Oates, S., Powell, M.B.: Identical relations in finite groups. J. Algebra **1**, 11–39 (1964)
15. Perkins, P.: Decision problems for equational theories of semigroups and general algebras. Ph.D. thesis, University of California, Berkeley (1966)

16. Perkins, P.: Bases for equational theories of semigroups. J. Algebra **11**, 298–314 (1969)
17. Sapir, M.V.: Problems of Burnside type and the finite basis property in varieties of semigroups. Izv. Akad. Nauk SSSR Ser. Mat. **51**, 319–340 (1987). [Russian (English trans.) Math. USSR-Izv. **30**, 295–314 (1987)]
18. Volkov, M.V.: The finite basis problem for finite semigroups, Sci. Math. Jpn. **53**,171–199 (2001). [A periodically updated version is avalaible under http://csseminar.imkn.urfu.ru/MATHJAP_revisited.pdf.]
19. Zimin, A.I.: Semigroups that are nilpotent in the sense of Mal'cev. Izv. Vyssh. Uchebn. Zaved. Mat., no.6, 23–29 (1980). [Russian]

Regular Elements in von Neumann Algebras

K.S.S. Nambooripad

Abstract The semigroup of all linear maps on a vector space is regular, but the semigroup of *continuous* linear maps on a Hilbert space is not, in general, regular; nor is the product of two regular elements regular. In this chapter, we show that in those types of von Neumann algebras of operators in which the lattice of projections is modular, the set of regular elements do form a (necessarily regular) semigroup. This is done using the construction of a regular biordered set (as defined in Nambooripad, Mem. Am. Math. Soc. 22:224, 1979, [9]) from a complemented modular lattice (as in Patijn, Semigroup Forum 21:205–220, 1980, [11]).

Keywords Rgular biordered set · $*$-Regular element

Throughout the following, H denotes a Hilbert space and $B(H)$ denotes the semigroup of bounded (continuous) linear maps of H to itself. Elements of $B(H)$ are called *operators* on H. The range and kernel of t in $B(H)$ are denoted by $\mathrm{ran}(t)$ and $\ker(t)$; that is for t in $B(H)$

$$\mathrm{ran}(t) = \{y \in H : \exists x \in H \;\; y = t(x)\} \quad \text{and} \quad \ker(t) = \{x \in H : t(x) = 0\}$$

By a *projection* of H, we mean a self-adjoint, idempotent operator on H; that is, an element p of $B(H)$ for which $p^* = p = p^2$. It is well known that $\mathrm{ran}(p)$ is a closed linear subspace of H for any projection p of H with $\ker(p) = \mathrm{ran}(p)^{\perp}$, and conversely, for each closed subspace A of H, there exists a projection p of H with $\mathrm{ran}(p) = A$ and $\ker(p) = A^{\perp}$. Thus there is a one-to-one correspondence between projections in $B(H)$ and closed linear subspaces of H. For each closed linear subspace A of H, the unique projection with range A is denoted as p_A and is called the projection on A.

An operator on H is called *regular*, if it is a regular element of the semigroup $B(H)$; that is, if there exists an operator t' on H such that $tt't = t$. If t' satisfies the equation $t'tt' = t'$ also, then it is called a generalized inverse of t. It is not difficult

K.S.S. Nambooripad (✉)
Department of Mathematics, University of Kerala, Kerala, India
e-mail: kssn267@gmail.com

P.G. Romeo et al. (eds.), *Semigroups, Algebras and Operator Theory*,
Springer Proceedings in Mathematics & Statistics 142,
DOI 10.1007/978-81-322-2488-4_3

to show that every linear map on a vector space has a generalized inverse (see [1], Sect. 2.3, Exercise 8). But in the case of an operator on a Hilbert space, it may happen that none of its generalized inverses are bounded, so that t is not regular as an element of $B(H)$. We can characterize the regular elements of $B(H)$ as follows (see [3]).

Proposition 1 *An element t of $B(H)$ is regular if and only if* $\operatorname{ran}(t)$ *is closed. In this case, there exists a unique generalized inverse $t^\dagger$ of t such that $tt^\dagger$ is the projection on* $\operatorname{ran}(t)$ *and $t^\dagger t$ is the projection on* $\ker(t)^\perp$. □

The generalized inverse $t^\dagger$ in the above result is called the *Moore–Penrose* inverse of t.

We next note that the map $t \mapsto t^*$, where t^* is the (Hilbert space) adjoint of t, is an involution on the semigroup $B(H)$ in the sense that $(t^*)^* = t$ for all t in $B(H)$ and $(st)^* = t^*s^*$ for all s, t in $B(H)$. Following [10], we define an element x of an involution semigroup S to be $*$-regular, if there exists an inverse $x^\dagger$ of x such that $(xx^\dagger)^* = xx^\dagger$ and $(x^\dagger x)^* = x^\dagger x$. Thus we have the following corollary:

Corollary 2 *An element t of $B(H)$ is $*$-regular if and only if it is regular.* □

Note that if p is a projection on H, then $\operatorname{ran}(p) = \ker(p)^\perp$ so that $p(x) = x$ for every $x \in \ker(p)^\perp$. This is generalized as follows: an operator v on H is called a *partial isometry*, if $||v(x)|| = ||x||$ for each x in $\ker(v)^\perp$. This can be characterized in several different ways (see [2], Proposition 4.38 and [4], Problem 98, Corollary 3).

Proposition 3 *For an operator v on H, the following are equivalent*

(i) v is a partial isometry
(ii) vv^* is a projection
(iii) v^*v is a projection
(iv) $vv^*v = v$ □

Now from (iv) above, we have

$$v^* = (vv^*v)^* = v^*v^{**}v^* = v^*vv^*$$

and this together with (iv) shows that v^* is an inverse of v. In view of (ii) and (iii), it follows that the adjoint v^* is in fact the Moore–Penrose inverse of v. Thus we can add the following characterization to the above list:

Proposition 4 *An operator v on H is a partial isometry if and only if v is a regular element of $B(H)$ for which $v^\dagger = v^*$.* □

If H is of finite dimension, then every linear subspace is closed and so the range of every operator is closed. Thus by Proposition 1, the semigroup $B(H)$ is regular. But if H is of infinite dimension, there are operators with nonclosed range (see [4], Sect. 41) and so $B(H)$ is not regular. Moreover, in this case, there are regular operators with nonregular products, so that the subset of $B(H)$ consisting only of regular operators does not form a semigroup. To see this, we give a necessary and sufficient condition for the product of two regular operators to be regular:

Proposition 5 *Let s and t be regular elements of $B(H)$. Then st is regular if and only if* $\ker(s) + \operatorname{ran}(t)$ *is closed in H.*

Proof We start by noting that for any two elements s, t in $B(H)$,

$$\ker(s) + \operatorname{ran}(t) = s^{-1}(\operatorname{ran}(st))$$

One of the inclusions is easily seen like this:

$$\ker(s) + \operatorname{ran}(t) \subseteq s^{-1}s(\ker(s) + \operatorname{ran}(t)) = s^{-1}(s(\operatorname{ran}(t)) = s^{-1}(\operatorname{ran}(st))$$

To prove the reverse inclusion, let $x \in s^{-1}(\operatorname{ran}(st))$, so that $s(x) = st(y)$ for some y in H. Hence $x - t(y) \in \ker(t)$ and so $x = (x - t(y)) + t(y) \in \ker(s) + \operatorname{ran}(t)$. Thus $s^{-1}(\operatorname{ran}(st)) \subseteq \ker(s) + \operatorname{ran}(t)$ also.

Now let s and t be regular elements of $B(H)$. If st is regular, then $\operatorname{ran}(st)$ is closed, by Proposition 1, and so $\ker(s) + \operatorname{ran}(t) = s^{-1}(\operatorname{ran}(st))$ is closed, since s is continuous. To prove the converse, first note that $\operatorname{ran}(s)$ is closed and hence complete, since s is regular, so that s is a continuous linear map from the complete normed space H onto the complete normed linear space $\operatorname{ran}(s)$ and so is an open map. Hence s is a quotient map onto $\operatorname{ran}(s)$ and $\operatorname{ran}(st) \subseteq \operatorname{ran}(s)$. If we now assume that $s^{-1}(\operatorname{ran}(st)) = \ker(s) + \operatorname{ran}(t)$ is closed in H, then it follows that $\operatorname{ran}(st)$ is closed in $\operatorname{ran}(s)$, since s is a quotient map. Again, since $\operatorname{ran}(s)$ is closed in H, it follows that $\operatorname{ran}(st)$ is closed in H. Thus st is regular. □

It can be shown that any Hilbert space of infinite dimension contains pairs of closed linear subspaces whose sum is not closed (see [5], Sect. 15) If A and B are two such subspaces, then by the above result, the product $p_{A^\perp} p_B$ is not regular. In other words, the set of regular elements of $B(H)$ is not a semigroup, if H is of infinite dimension.

This result can also be used to show how the regularity of a product in the semigroup $B(H)$ is linked to the lattice of closed linear subspaces of H. It is easily seen that the set of all subspaces of a vector space is a lattice under set inclusion, with meet and join defined by

$$A \wedge B = A \cap B \qquad A \vee B = A + B$$

and that it satisfies the *modular identity*

$$A \leq C \implies A \vee (B \wedge C) = (A \vee B) \wedge C$$

If H is a Hilbert space, then the set of all *closed* linear subspaces is not a sublattice of the above, since the sum of two closed linear subspaces may not be closed in general. However, it is indeed a lattice with meet and join defined by

$$A \wedge B = A \cap B \qquad A \vee B = \operatorname{cl}(A + B)$$

where cl denotes closure in H. But this lattice is modular if and only if H is of finite dimension, the lack of modularity being a consequence of the existence of closed linear subspaces with nonclosed sum (see [4], Problem 9). This is seen in a better perspective by localizing the notion of modularity: two closed linear subspaces A, B of H is said to be a *modular pair* if

$$X \leq B \implies X \vee (A \wedge B) = (X \vee A) \wedge B$$

It can be shown that A, B form a modular pair if and only if $A + B$ is closed ([7], Theorem III-6 and Theorem III-13) Thus our result on the regularity of a product can be rephrased as follows:

Proposition 6 *Let s and t be regular elements in $B(H)$. Then st is regular if and only if* $\ker(s)$ *and* $\operatorname{ran}(t)$ *form a modular pair in the lattice of closed subspaces of H* □

It may also be noted that the one-to-one correspondence $A \mapsto p_A$, between closed linear subspaces of H and projections on them, induces a lattice structure on the set of projections of H. We call this the *projection lattice* of H and denote it as $P(H)$.

We now pass on to the definition of a von Neumann algebra. This can be done in several equivalent ways, and we choose the one which uses only the algebraic properties of $B(H)$ (see [12]). For this, we define the *commutant* of a subset S of $B(H)$, denoted S', by

$$S' = \{t \in B(H) : st = ts \ \forall s \in S\}$$

Note that a subset S of $B(H)$ is called self-adjoint, if for each t in S, we also have t^* in S.

Definition 7 A self-adjoint subalgebra A of $B(H)$ is called a von Neumann algebra if $A'' = A$.

An interesting feature of a von Neumann algebra A is that, as remarked in [12], just about any canonical construction applied to the elements of A results in an element of A itself.

For example, consider the following decomposition of an operator: it can be proved that for any operator t on H, there is a unique pair of operators s and v where s is a positive operator (meaning $\langle t(x), x\rangle \geq 0$ for every x in H) and v is a partial isometry with $\ker(s) = \ker(v)$ such that $t = vs$ (see [2], Proposition 4.39). This is called the *polar decomposition* of an operator. If t is in a von Neumann algebra A and $t = vs$ is the polar decomposition of t, then both v and s are in A ([12], Corollary 0.4.9). As another example, we show that if $t \in A$, then its Moore–Penrose inverse $t^\dagger$ is also in A. For this we make use of the following result proved in [12] (Scholium 0.4.8). Note that an operator u on H is called *unitary* if $uu^* = u^*u = 1$, the identity operator on H.

Lemma 8 *Let A be a von Neumann algebra of operators on H and let t be an operator on H. Then $t \in A$ if and only if $utu^* = t$ for every unitary operator u in the commutant A' of A.* □

Proposition 9 *Let t be a regular element of $B(H)$ and let A be a subset of $B(H)$ which is a von Neumann algebra. If t is in A, then the Moore–Penrose inverse $t^\dagger$ of t is also in A.*

Proof Let $t \in A$. We first prove that the projections $tt^\dagger$ and $t^\dagger t$ are in A. Let $t = vs$ be the polar decomposition of t, so that $v \in A$. Since A is self-adjoint, we have $v^* \in A$ and hence vv^* and v^*v are in A. Moreover, vv^* is the projection on $\operatorname{ran}(v) = \operatorname{ran}(t)$ and v^*v is the projection on $\ker(v)^\perp = \ker(t)^\perp$ (see Proposition 6.1.1 and Theorem 6.1.2 of [6]). By definition of $t^\dagger$, it follows that $tt^\dagger = vv^*$ and $t^\dagger t = v^*v$. Thus $tt^\dagger$ and $t^\dagger t$ are in A.

To show that $t^\dagger$ itself is in A, we make use of the lemma above. Let u be a unitary operator in A'. $ut^\dagger u^* = t^\dagger$, we start by proving that $t' = ut^\dagger u^*$ is an inverse of t in $B(H)$. Since $t \in A$ and $u \in A'$, we have $ut = tu$. Again, A' is self-adjoint, since A is ([12], Proposition 0.4.1) so that $u^* \in A'$ and so $u^*t = tu^*$. Thus

$$tt't = t(ut^\dagger u^*)t = (tu)t^\dagger(tu^*) = (ut)t^\dagger(tu^*) = u(tt^\dagger t)u^* = utu^* = tuu^* = t$$

since u is unitary. Again, since $u^*tu = tu^*u = t$,

$$t'tt' = (ut^\dagger u^*)t(ut^\dagger u^*) = (ut^\dagger)(u^*tu)(t^\dagger u^*) = u(t^\dagger tt^\dagger)u^* = ut^\dagger u^* = t'$$

Thus t' is an inverse of t. Also,

$$tt' = t(ut^\dagger u^*) = u(tt^\dagger)u^* = tt^\dagger$$

and

$$t't = (ut^\dagger u^*)t = u(t^\dagger t)u^* = t^\dagger t$$

since $tt^\dagger$ and $t^\dagger t$ are in A. So, t' is an inverse of t in $B(H)$ with tt' and $t't$ projections, so that $t' = t^\dagger$, by the uniqueness of the Moore–Penrose inverse. Thus $ut^\dagger u^{**} = t^\dagger$ for every unitary operator in A' and hence $t^\dagger \in A$, by the lemma. □

This result immediately leads to the following generalization of Corollary 2.

Corollary 10 *An element t of a von Neumann algebra is $*$-regular if and only if it is regular.* □

Another property of a von Neumann algebra A is that the set $P(A)$ of projections in it forms a complete sublattice of $P(H)$. In [8], certain kinds of von Neumann algebras are classified into different *types*, on the basis of a real valued, positive function on the projection lattice. A von Neumann algebra A is called a *factor* if

$$A \cap A' = \{\lambda 1 : \lambda \in \mathbb{C}\}$$

that is, a factor is a von Neumann algebra A in which the only operators in A commuting with all operators in A are scalar multiples of the identity operator. In [8], factors are classified into five types named I_n, I_∞, II_1, II_∞, III (see [12] for a concise description of these ideas).

Our interest in this scheme is that for factors of type I_n and II_1, the projection lattice is modular. So, for a factor A of these types, any two projections form a modular pair in the projection lattice $P(A)$ and so by Proposition 6, the product of any two regular elements of A is again regular. Thus we have the following:

Proposition 11 *If A is a factor of type* I_n *or* II_1*, then the set of regular operators in A forms a regular subsemigroup of A.*

Another approach to this idea is via the notion of *biordered sets*, introduced in [9] and the construction of a biordered set from a modular lattice, given in [11]. A biordered set is defined as a set with a partial binary operation and the set of idempotents of a regular semigroups is characterized as a special kind of biordered set termed *regular biordered set* (see [9] for the details). In [11], it is shown if L is a complemented modular lattice, then the set $E(L)$ of pairs of complementary elements of L can be turned into a regular biordered set with a suitably defined partial binary operation. Thus it follows that for a von Neumann algebra A whose projection lattice is modular, the set of regular operators form a semigroup. Moreover, this approach also gives explicit expressions for the Moore–Penrose inverse of the product of regular elements:

Proposition 12 *Let s and t regular operators in a von Neumann algebra of type* I_n *or* II_1 *and let p and q be the projections on* $\ker(st)^\perp$ *and* $\operatorname{ran}(s)$*, respectively. If h is the idempotent operator of H with*

$$\operatorname{ran}(h) = (\ker(s)^\perp + \operatorname{ran}(t)^\perp) \cap \operatorname{ran}(t)$$
$$\ker(h) = (\ker(s)^\perp \cap \operatorname{ran}(t)^\perp) + \ker(s)$$

then $(st)^\dagger = pt^\dagger hs^\dagger q$. □

References

1. Clifford, A.H., Preston, G.B.: The Algebraic Theory of Semigroups, vol. I. American Mathematical Society, Providence (1961)
2. Douglas, R.G.: Banach Algebra Techniques in Operator Theory. Springer, New York (1998)
3. Groetsch, C.W.: Generalized Inverses of Linear Operators. Marcel Dekker, New York (1977)
4. Halmos, P.R.: A Hilbert Space Problem Book. Springer, New York (1982)
5. Halmos, P.R.: Introduction to Hilbert Space. Chelsea, Providence (1998)
6. Kadison, R.V., Ringrose, J.R.: Fundamentals of Operator Algebras. American Mathematical Society, Providence (1997)
7. Mackey, G.W.: On infinite-dimensional linear spaces. Trans. Am. Math. Soc. **57**, 155–207 (1945)

8. Murray, F.J., von Neumann, J.: On rings of operators. Ann. Math. **37**, 116–229 (1936)
9. Nambooripad, K.S.S.: Structure of regular semigroups-I. Mem. Am. Math. Soc. **22**, 224 (1979)
10. Nambooripad, K.S.S., Pastijn, F.J.C.M.: Regular involution semigroups. Semigroups Colloq. Math. Soc. Janos Bolyayi **39**, 199–249 (1981)
11. Patijn, F.: Biordered sets and complemented modular lattices. Semigroup Forum **21**, 205–220 (1980)
12. Sunder, V.S.: An Invitation to von Neumann Algebras. Springer, New York (1986)
13. von Neumann, J.: Continuous Geometry. Princeton University Press, Princeton (1998)

LeftRight Clifford Semigroups

M.K. Sen

Abstract Clifford semigroups are certain interesting class semigroups and looking for regular semigroups close to this is natural. Here we discuss the leftright Clifford semigroups.

Keywords LC-semigroup · Clifford semigroup · LR-C semigroup

1 Introduction

Let S be a semigroup and denote its set of idempotents by E_S. Recall that S is regular if for each $a \in S$ there exists $x \in S$ such that $a = axa$. As usual, Greens relations are denoted by $\mathcal{L}$, $\mathcal{R}$, $\mathcal{H}$, $\mathcal{D}$, and $\mathcal{J}$. An element of a semigroup S is completely regular, if there exists an element $x \in S$ such that $axa = a$ and $ax = xa$. A semigroup S is completely regular semigroup if all its elements are completely regular. A completely regular semigroup is a semigroup that is a union of groups. This class of regular semigroups and its subclasses have been studied by many authors (see M. Petrich and N.R. Reilly [7]).

A semigroup S is said to be an inverse semigroup if, for every $a \in S$, there is a unique $b \in S$ (called the inverse of a) such that $aba = a$ and $bab = b$. This is equivalent to the condition that semigroup S is regular and that the idempotents of S commute. It follows that every inverse semigroup which is completely regular, i.e., union of groups is a semilattice of groups, and conversely. Any such semigroup is now known as Clifford semigroup. Such semigroup can be constructed as follows. Let Y be a semilattice, and to each element α of Y assign a group G_α such that G_α and G_β are disjoint if $\alpha \neq \beta$. To each pair of elements α, β of Y such that $\alpha \geq \beta$, assign a homomorphism $\phi_{\alpha,\beta}$: $G_\alpha \to G_\beta$ such that:

(1.1) $\alpha \geq \beta \geq \gamma$ implies $\phi_{\alpha,\beta}\phi_{\beta,\gamma} = \phi_{\alpha,\gamma}$,
(1.2) $\phi_{\alpha,\alpha}$ is the identity automorphism of G_α.

M.K. Sen (✉)
Department of Pure Mathematics, University of Kolkata, Kolkata, India
e-mail: senmk6@yahoo.com

P.G. Romeo et al. (eds.), *Semigroups, Algebras and Operator Theory*,
Springer Proceedings in Mathematics & Statistics 142,
DOI 10.1007/978-81-322-2488-4_4

Let $S = \bigcup\{G_\alpha | \alpha \in Y\}$ and define the product ab of two elements a, b of S as follows: if $a \in G_\alpha$ and $b \in G_\beta$ then $ab = (a\phi_{\alpha,\alpha\beta})(b\phi_{\beta,\alpha\beta})$. Here $\alpha\beta$ is the product of α and β in Y. We call $\{\phi_{\alpha,\beta} \mid \alpha \geq \beta\}$, the set of connecting homomorphisms of S.

2 Orthogroups

A semigroup S is called orthodox if it is regular ($a \in aSa$ for all a in S) and if the set E_S of idempotents of S is a subsemigroup of S. An orthodox semigroup is called an orthogroup if it is also a completely regular semigroup. It is well known that an orthogroup can be expressed as a semilattice of rectangular groups. Recall that a rectangular group is a semigroup isomorphic to the direct product $L \times G \times R$ where L is a left and R is a right zero semigroup, and G is a group. The structure of orthogroups has been studied by many authors, and in particular, Petrich [6] in 1987 described an elegant method for the construction of such semigroups.

Let Y be a semilattice. For every $\alpha \in Y$, let $S_\alpha = I_\alpha \times G_\alpha \times \Lambda_\alpha$, where I_α is a left zero semigroup, G_α is a group, and Λ_α is a right zero semigroup, and assume that $S_\alpha \cap S_\beta = \phi$ if $\alpha \neq \beta$. For each $\alpha \in Y$, fix an element in $I_\alpha \cap \Lambda_\alpha$.

Let $\langle , \rangle : S_\alpha \times I_\beta \to I_\beta$, $[,] : \Lambda_\beta \times S_\alpha \to \Lambda_\beta$ be two functions defined whenever $\alpha \geq \beta$. Let G be a semilattice Y of groups G_α in which multiplication is denoted by juxtaposition. Assume that for all $a = (i, g, \lambda) \in S_\alpha$ and $b = (j, h, \mu) \in S_\beta$, the following conditions hold.
(A) If $k \in I_\alpha$ and $\nu \in \Lambda_\alpha$, then $\langle a, k\rangle = i$, and $[\nu, a] = \lambda$. On $S = \bigcup_{\alpha\in Y} S_\alpha$ define a multiplication by $aob = (\langle a, \langle b, \alpha\beta\rangle\rangle, gh, [[\alpha\beta, a], b])$.
(B) If $\gamma \leq \alpha\beta$, $k \in I_\gamma$, $\nu \in \Lambda_\gamma$, then $\langle a, \langle b, k\rangle\rangle = \langle aob, k\rangle$, $[[\nu, a], b] = [v, aob]$.
Then S is an orthogroup such that $S/\mathcal{D} \cong Y$ and whose multiplication restricted to each S_α coincides with the given multiplication. Conversely, every orthogroup is isomorphic to one so constructed.

Petrich and Reilly [7] have discussed some special orthogroups, that is, the $\mathcal{C}$ orthogroup if its set of idempotents forms a $\mathcal{C}$ band. The most well-known $\mathcal{C}$ bands are tabled in the text of Howie [8], except semilattice and rectangular band, as follows:

(1) regular band: the band satisfying the identity $efge = efege$,
(2) left regular band: the band satisfying the identity $ef = efe$, and right regular band: the band satisfying the identity $fe = efe$,
(3) normal band: the band satisfying the identity $efge = egfe$,
(4) left normal band: the band satisfying the identity $efg = egf$, and right normal band: the band satisfying the identity $fge = gfe$.

Among the above classes of $\mathcal{C}$ orthogroups, the class of left regular orthogroups and the class of right regular orthogroups are two important proper subclasses of the class of regular orthogroups, and in particular, their intersection is the class of Clifford semigroups.

3 $\mathcal{LC}$-Semigroups

Due to the rich structure of Clifford semigroup, it is natural to search for classes of regular semigroups close to Clifford semigroups. In [5], Petrich, and in [13], Zhu et al. left [right]Clifford semigroups as generalization of Clifford semigroups. A left [right]Clifford semigroup is a regular semigroup S satisfying $eS \subseteq Se$ [resp. $Se \subseteq eS$], for all $e \in E_S$. For the sake of convenience, we call the left Clifford semigroups as just $\mathcal{LC}$-semigroups. We discuss some basic properties of $\mathcal{LC}$-semigroups.

Lemma 1 *Let S be a regular semigroup with the set of idempotents E. Then S is an $\mathcal{LC}$-semigroup if and only if for any $e \in E$ and $a \in S$, $eae = ea$.*

Lemma 2 *A semigroup S is $\mathcal{LC}$ if and only if S is completely regular and the set E_S of idempotents of S is a left regular band (that is, S is a left regular orthogroup).*

Lemma 3 *Let S be an $\mathcal{LC}$ and $s \in S$. Then $aa' = aa''$ for all $a', a'' \in V(A)$, that is, S is a right inverse semigroup.*

Note. The converse of the above result is not true. For instance, the bicyclic semigroup is a right inverse semigroup but not $\mathcal{LC}$.

In the following we list equivalent conditions obtained by [13]:

1. S is left Clifford.
2. S is regular and $\mathcal{L} = \mathcal{J}$ is a semilattice congruence.
3. S is a semilattice of left groups.
4. S is regular such that each idempotent e of S lies in the center of eS.
5. S is regular with $aS \subseteq Sa$, for all $a \in S$.
6. S is regular and $\mathcal{D}^S \bigcap (Es \times Es) = \mathcal{L}^{Es}$.

Theorem 4 *Let S be a semigroup with set of idempotents E_S. Then S is a strong semilattice of left groups if and only if S is the left normal band of groups and E_S is a band.*

4 A Structure Theorem

In [10] Sen et al. introduced the concept of left quasi-direct product of semigroups. Let Y be a semilattice and T a Clifford semigroup with Y as the semilattice of its idempotents. Following Kimura [9], in [10] we find construction of left regular band $B = \bigcup_{\alpha \in Y} B_\alpha$, where each B_α is a left zero band, for all $\alpha \in Y$. Let $B \oplus T = \{(e, \gamma) \mid \gamma \in T, e^2 = e \in B_{\gamma_0}\}$. Define a mapping ϕ from T into the endomorphism semigroup End(B) of B by

$$\phi : \gamma \to \gamma\phi = \sigma_\gamma, \text{ where } \gamma \in T \text{ and } \sigma_\gamma \in \text{End}(B)$$

such that the following conditions are satisfied:

(P_1) For any $\gamma \in T$ and $\alpha \in Y$, $B_\alpha\sigma_\gamma \subseteq B_{(\alpha\gamma)^0}$; and if $\gamma \in Y$, then there exists $f^2 = f \in B_\gamma$ such that $e\sigma_\gamma = fe$, for any $e^2 = e \in B$.

(P_2) For any $\gamma, w \in T$, $f \in B_{(\gamma w)^0}$, we have $\sigma_w\sigma_\gamma\delta_f = \sigma_{\gamma w}\delta_f$, where δ_f is an inner endomorphism on B (that is, $e\delta_f = fef = fe$) for any $e \in B$.

Now, define a *multiplication* $*$ on $B \oplus T$ by

$$(e, w) * (f, \tau) = \Big(e(f\sigma_w),\ w\tau\Big)$$

for any $(e, w), (f, \tau) \in B \oplus T$. Then $B \oplus T$ forms a semigroup under the multiplication "$*$". This semigroup $(B \oplus T, *)$, is called a left quasi-direct product of B and T which is determined by the mapping ϕ, denoted by $B \oplus_\phi T$. Right quasi-direct product can be likewisely defined. With the above definition of left quasi-direct product of semigroups, it was proved in [10], the following construction theorem for $\mathcal{LC}$-semigroups.

Theorem 5 *Let T be a Clifford semigroup whose set of idempotents forms a semilattice Y. Let $B = \bigcup_{\alpha \in Y} B_\alpha$ be a left regular band. Then the left quasi-direct product $B \oplus_\phi T$ of B and T is an $\mathcal{LC}$-semigroup. Conversely, every $\mathcal{LC}$-semigroup S is isomorphic to a left quasi-direct product of a left regular band and a Clifford semigroup.*

5 $\mathcal{LR}$-$\mathcal{C}$ Semigroup

In [11] Sen et al. introduced $\mathcal{LR}$-$\mathcal{C}$ semigroup.

Definition 6 A semigroup S is called leftright Clifford, abbreviated to $\mathcal{LR}$-$\mathcal{C}$, if S is regular and for all idempotents $e \in E_S$, $eS \subseteq Se$ or $Se \subseteq eS$.

Clearly, all left and all right Clifford semigroups are $\mathcal{LR}$-$\mathcal{C}$. But the converse may fail.

Example 7 Let $S = \{(a, b) \in R \times R \mid ab = 0\}$ with multiplication $(a, b)(c, d) = (|a|c, b|d|)$, with respect to which S becomes a regular semigroup.

$E_S = \Big\{(1, 0), (-1, 0), (0, 0), (0, 1), (0, -1)\Big\}$.
Also $S(1, 0) \subseteq (1, 0)S$,
$S(-1, 0) \subseteq (-1, 0)S$,
$(0, 1)S \subseteq S(0, 1)$,
and $(0, -1)S \subseteq S(0, -1)$.

Hence S is an $\mathcal{LR}$-$\mathcal{C}$ semigroup but S is neither a left Clifford semigroup nor a right Clifford semigroup as $(1, 0)S$ is not a subset of $S(1, 0)$ and $S(0, 1)$ is not a subset of $(0, 1)S$.

Lemma 8 *Let S be an $\mathcal{LR}$-$\mathcal{C}$ semigroup. Then for $e, f \in E_S$, either $efe = ef$ or $efe = fe$.*

Proof For $e \in E_S$, either $eS \subseteq Se$ or $Se \subseteq eS$. If $eS \subseteq Se$, then for any $f \in E_S$, $ef = te$ for some $t \in S$. Then $efe = tee = te = ef$. Again, if $Se \subseteq eS$, $fe = et_1$ for some $t_1 \in S$ so that $efe = eet_1 = et_1 = fe$. ■

Definition 9 A band B is called an $\mathcal{LR}$ regular band if for $e, f \in E_S$, either $efe = ef$ or $efe = fe$.

A left(right) regular band is clearly an $\mathcal{LR}$ regular band but the converse is not necessarily true, the following is an example.

Example 10 Let $B = \{e_1, e_2, f_1, f_2, g\}$ with the following table

$\cdot$	e_1	e_2	f_1	f_2	g
e_1	e_1	e_2	g	g	g
e_2	e_1	e_2	g	g	g
f_1	g	g	f_1	f_1	g
f_2	g	g	f_2	f_2	g
g	g	g	g	g	g

One can check that B is an $\mathcal{LR}$ regular band but B is neither a left regular band nor a right regular band. This is because that $e_1e_2 \neq e_1e_2e_1$ and $f_1f_2 \neq f_2f_1f_2$.

We know that a semigroup S is completely semigroup if and only if $a \in aSa^2$, for all $a \in S$.

Theorem 11 *Let S be an $\mathcal{LR}$-$\mathcal{C}$ semigroup. Then S is completely regular.*

Proof Let a be an element of S and x be an inverse of a in S. Let $e = xa$ and $eS \subseteq Se$. Then $x = xax = ex = x_1e$ for some $x_1 \in S$. Hence $x = x_1xa$ which implies that $x^2 = x_1xax = x_1x$ so that $x = x_1xa = x^2a$ and $ax = ax^2a$. Suppose now $Se \subseteq eS$. Then $a = axa = ae = ea_1$ for some a_1 in S. Now $a = ea_1 = xaa_1$ which implies that $a^2 = axaa_1 = aa_1$ and so $a = xa^2$. Hence $ax = xa^2x$. Then either $ax = ax^2a$ or $ax = xa^2x$. If $ax = ax^2a$, then $a = ax^2a^2 \in aSa^2$. Also if $ax = xa^2x$, then $a = axa = xa^2xa = xa^2$ so that $a = axa = axxa^2 = ax^2a^2 \in aSa^2$. Thus in both cases, $a \in aSa^2$ which implies that S is completely regular. ■

The following is an example of a completely regular semigroup which is not an $\mathcal{LR}$-$\mathcal{C}$ semigroup.

Example 12 Let $\mathbb{Q}^+$ be the set of positive rational numbers and

$$S = \left\{ \begin{pmatrix} a & b \\ xa & xb \end{pmatrix} : a, b, x \in \mathbb{Q}^+ \right\}.$$

Then S is a completely regular semigroup under usual multiplication of matrices.

Indeed, $\begin{pmatrix} a & b \\ xa & xb \end{pmatrix}\begin{pmatrix} c & d \\ yc & yd \end{pmatrix} = (a+yb)\begin{pmatrix} c & d \\ xc & xd \end{pmatrix}$. Also $\begin{pmatrix} a & b \\ xa & xb \end{pmatrix} \in E_S$ if and only if $a + xb = 1$. But S is not an LR-C semigroup, since for $E = \begin{pmatrix} \frac{1}{2} & 1 \\ \frac{1}{4} & \frac{1}{2} \end{pmatrix} \in E_S$ and $F = \begin{pmatrix} \frac{1}{3} & 2 \\ \frac{1}{9} & \frac{2}{3} \end{pmatrix} \in S$, we have $EF = \frac{5}{6}\begin{pmatrix} \frac{1}{3} & 2 \\ \frac{1}{6} & 1 \end{pmatrix} \notin SE$. In fact, for any $A = \begin{pmatrix} a & b \\ xa & xb \end{pmatrix} \in S$, $AE = (a+\frac{b}{2})\begin{pmatrix} \frac{1}{2} & 1 \\ \frac{x}{2} & x \end{pmatrix}$. Then $EF = AE$ implies $\frac{1}{2}(a+\frac{b}{2}) = \frac{5}{18}$ and $(a+\frac{b}{2}) = \frac{5}{3}$ which is impossible for any $a, b \in \mathbb{Q}^+$. Also $FE = \frac{4}{3}\begin{pmatrix} \frac{1}{2} & 1 \\ \frac{1}{6} & \frac{1}{3} \end{pmatrix} \notin ES$. For, if $A = \begin{pmatrix} a & b \\ xa & xb \end{pmatrix} \in S$, then $EA = (\frac{1}{2}+x)\begin{pmatrix} a & b \\ \frac{a}{2} & \frac{b}{2} \end{pmatrix}$. Then $EA = FE$ implies $(\frac{1}{2}+x)a = \frac{2}{3}$ and $(\frac{1}{2}+x)\,\frac{1}{2}a = \frac{2}{9}$, which is again impossible for any $a, x \in \mathbb{Q}^+$. Thus S is not an $\mathcal{LR}$-$\mathcal{C}$ semigroup.

Theorem 13 *Let S be a regular semigroup. Then the following statements are equivalent:*

(i) *S is an* $\mathcal{LR}$-$\mathcal{C}$ *semigroup.*
(ii) *For each* $a \in S$, *either* $aS \subseteq Sa$ *or* $Sa \subseteq aS$.
(iii) *For each* $a \in S$, *either* $aS \subseteq Sa$ *or there exists an inverse* a^* *of a in S such that* $Sa \subseteq a^*aS$.
(iv) *For each* $a \in S$, *either* $Sa \subseteq aS$ *or there exists an inverse* a^* *of a in S such that* $aS \subseteq Saa^*$.

Theorem 14 *A semigroup S is an* $\mathcal{LR}$-$\mathcal{C}$ *semigroup if and only if S is completely regular and for* $e, f \in E_S$, *either* $efe = ef$, *or* $efe = fe$ *(that is,* E_S *is* $\mathcal{LR}$ *regular band).*

Proof Let $e \in E_S$ and $x \in S$. Then S being completely regular, $ex = extex = exext$, for some $t \in S$. Suppose $efe = ef$, for all $f \in E_S$, then $ex = exe(ext) = exe(ext)e \in Se$ so that $eS \subseteq Se$. Similarly, if $efe = fe$ for all $f \in E_S$, then $Se \subseteq eS$. Hence S is an $\mathcal{LR}$-$\mathcal{C}$ semigroup. ■

Recall that a regular semigroup S is orthodox if E_S is a subsemigroup of S.

Theorem 15 *Let S be an* $\mathcal{LR}$-$\mathcal{C}$ *semigroup. Then*

(i) *S is orthodox,*
(ii) *for every* $a \in S$, *either* $aV(a) \subseteq V(a)a$ *or* $V(a)a \subseteq aV(a)$.

The next example describes an orthodox semigroup which is not an $\mathcal{LR}$-$\mathcal{C}$ semigroup.

Example 16 We consider the inverse semigroup $\mathcal{I}(X)$ of all one-one partial mappings of the set $X = \{1, 2, 3\}$. $\mathcal{I}(X)$ is an orthodox semigroup. It is to be noted that an element α of $\mathcal{I}(X)$ is idempotent if and only if α is the identity mapping of a subset A of X onto itself. We show that $\mathcal{I}(X)$ is not an $\mathcal{LR}$-$\mathcal{C}$. For this, we consider the elements $e = \begin{pmatrix} 1 \\ 1 \end{pmatrix}$ and $f = \begin{pmatrix} 1\,2 \\ 2\,1 \end{pmatrix}$ in $\mathcal{I}(X)$, where e is an idempotent. Then $ef = \begin{pmatrix} 2 \\ 1 \end{pmatrix} \notin \mathcal{I}(X)e$ and $fe = \begin{pmatrix} 1 \\ 2 \end{pmatrix} \notin e\mathcal{I}(X)$. This implies that $e\mathcal{I}(X)$ is not a subset of $\mathcal{I}(X)e$ and $\mathcal{I}(X)e$ is not a subset of $e\mathcal{I}(X)$. Hence $\mathcal{I}(X)$ is not an $\mathcal{LR}$-$\mathcal{C}$ semigroup.

6 Structure of $\mathcal{LR}$-$\mathcal{C}$ Semigroups

A semigroup S is called a quasi-inverse semigroup, if it is orthodox and for any $a, x, y, a \in E_S$, $axya = axaya$ (that is, E_S is a regular band).

Let S be an $\mathcal{LR}$-$\mathcal{C}$ semigroup. Then for all $e \in E_S$, either $efe = ef$, for all $f \in E_S$ or $efe = fe$, for all $f \in E_S$. Now let $a, x, y \in E_S$. Suppose $afa = af$, for all $f \in E_S$. Then $a(xy)a = (ax)y = axay$ so that $axya = axaya$. This also holds in case $afa = fa$, for all $f \in E_S$. Hence E_S is a regular band. Also from the Theorem 15 it follows that S is orthodox. Hence S is a quasi-inverse semigroup.

The converse of the above result may not be true. In fact, the inverse semigroup $\mathcal{I}(X)$ of Example 16 is a quasi-inverse semigroup which is not an $\mathcal{LR}$-$\mathcal{C}$ semigroup.

Let S be a quasi-inverse semigroup. Then S is an orthodox semigroup. The least inverse semigroup congruence σ on S is given as follows:

$a\sigma b$ if and only if $V(a) = V(b)$, where $V(x)$ denotes the set of all inverses of x.

Now, a relation η_1 on S is defined as follows:

$a\eta_1 b$ if and only if

(1) $\bar{a} = \bar{b}$ (where $\bar{x}$ denotes the σ-class containing $x \in S$),
(2) there exists an inverse a^* of a and an inverse b^* of b such that $aea^*\, beb^* = beb^*$ and $beb^*\, aea^* = aea^*$, for all $e \in E_S$.

By [12], η_1 is a congruence on S and S/η_1 is a left inverse semigroup.

Dually, a relation η_2 on S is defined by

$a\eta_2 b$ if and only if

(1) $\bar{a} = \bar{b}$ and
(2) there exists an inverse a^* of a and an inverse b^* of b such that $a^*ea\, b^*eb = a^*ea$ and $b^*eb\, a^*ea = b^*eb$, for all idempotents $e \in E_S$.

Then η_2 is a congruence on S and S/η_2 is a right inverse semigroup. According to Yamada [12], a regular semigroup is a quasi-inverse semigroup if and only if it is isomorphic to a subdirect product of a left inverse semigroup and a right inverse semigroup.

We now show that a quasi-inverse semigroup may be characterized as a special type of subdirect product, namely, spined product. The direct product of a family of semigroups $\{S_\alpha\}_{\alpha\in A}$ is denoted by $\prod_{\alpha\in A} S_\alpha$. Let $S = \prod_{\alpha\in A} S_\alpha$ and π_α denote the projection homomorphism $\pi_\alpha : S \to S_\alpha$. Any semigroup S isomorphic to a subsemigroup P of S such that $P\pi_\alpha = S_\alpha$, for all $\alpha \in A$ is called a subdirect product of the semigroups $S_\alpha, \alpha \in A$. Let T be a semigroup and suppose that for some family of semigroups $\{S_\alpha\}_{\alpha\in A}$ and each $\alpha \in A$, there exists an epimorphism $\varphi_\alpha : S_\alpha \to T$. Then

$$S = \left\{ (a_\alpha) \in \prod_{\alpha\in A} S_\alpha | a_\alpha\varphi_\alpha = a_\beta\varphi_\beta, \alpha, \beta \in A \right\}$$

is called the spined product of semigroups S_α over T (and the epimorphisms φ_α, $\alpha \in A$). It follows easily that the spined product is a subdirect product of semigroups S_α. If it is clear from the context what are the epimorphisms are, we say only that S is a spined product of semigroups S_α.

Theorem 17 *A quasi-inverse semigroup is isomorphic to a spined product of a left inverse semigroup and a right inverse semigroup with respect to an inverse semigroup which is the greatest inverse semigroup homomorphic image of S.*

Proof Let S be a quasi-inverse semigroup. By [12], S is isomorphic to a subdirect product of S/η_1 and S/η_2 where the isomorphism $\phi : S \longrightarrow S/\eta_1 \times S/\eta_2$ is given by $a\phi = (\tilde{a}, \tilde{\tilde{a}})$, $a \in S$ where $\tilde{a}, \tilde{\tilde{a}}$ denote the η_1-class and η_2-class containing a, respectively. Let $\psi : S/\eta_1 \longrightarrow S/\sigma$ and $\xi : S/\eta_2 \longrightarrow S/\sigma$ be defined by $\tilde{a}\psi = \bar{a}$ and $\tilde{\tilde{a}}\xi = \bar{a}$, where $\bar{a}$ is the σ-class containing a. Then it is easy to see that ψ and ξ are epimorphisms. Thus $T = \{(\tilde{a}, \tilde{\tilde{b}}) \in S/\eta_1 \times S/\eta_2 : a\sigma b\}$ is a spined product of S/η_1 and S/η_2 with respect to S/σ. Now define $\theta : S \longrightarrow T$ by $a\theta = (\tilde{a}, \tilde{\tilde{a}})$. Clearly, θ is a monomorphism. Let $(\tilde{a}, \tilde{\tilde{b}}) \in T$. Then $a\sigma b$ so that $V(a) = V(b)$ and $a = ab^*a = a(b^*bb^*)a = a(b^*b)(b^*a)\,\eta_1\, a(b^*b)(b^*a)(b^*b) = ab^*ab^*b = ab^*b = c$ (say) where $b^* \in V(b)$. Also $c = ab^*b = ab^*(bb^*b) = (ab^*)(bb^*)b\ \eta_2\ (bb^*)(ab^*)(bb^*)b = b(b^*ab^*)b = bb^*b = b$. Therefore, $c = ab^*b \in S$ such that $\tilde{c} = \tilde{a}, \tilde{\tilde{c}} = \tilde{\tilde{b}}$. Thus $c\theta = (\tilde{a}, \tilde{\tilde{b}})$ and θ is an isomorphism. ■

Now suppose S is an $\mathcal{LR}$-$\mathcal{C}$ semigroup. Then by Theorem 11, S is completely regular. Also a homomorphic image of a completely regular semigroup is completely regular. Consequently, the left inverse semigroup S/η_1 becomes a left Clifford semigroup. Similarly, S/η_2 is a right Clifford semigroup and S/σ becomes a Clifford semigroup. Thus we have the following:

Corollary 18 *An $\mathcal{LR}$-$\mathcal{C}$ semigroup S is isomorphic to a spined product of a left Clifford semigroup and a right Clifford semigroup with respect to a Clifford semigroup.*

Example 19 Let R be the set of real numbers. For any $a, b \in R$, we define $a \bullet b = a|b|$, $a \circ b = |a|b$. Then $L = (R, \bullet)$ is a left Clifford semigroup and $R = (R, \circ)$ is a right Clifford semigroup. Let C be the semigroup of nonnegative real numbers with usual multiplication. Then C is a Clifford semigroup. We define the mappings $\theta : L \longrightarrow C$ by $a\theta = |a|$ and $\phi : R \longrightarrow C$ by $a\phi = |a|$. It is easy to verify that θ and ϕ are epimorphisms. Let $T = \Big\{(a, b) \in L \times R : a\theta = b\phi\Big\}$ be the spined product of L and R with respect to C. Now $T = \Big\{(a, a), (a, -a) : a \in R\Big\}$ is not an $\mathcal{LR}$-$\mathcal{C}$ semigroup. Indeed, for the element $(1, 1) \in E_T$, we have $(1, 1)(-1, -1)(1, 1) = (1, 1)$ whereas $(1, 1)(-1, -1) = (1, -1)$ and $(-1, -1)(1, 1) = (-1, 1)$.

Theorem 20 *Let S be a semigroup. The following conditions are equivalent:*

(1) S is an $\mathcal{LR}$-$\mathcal{C}$ semigroup.
(2) S is isomorphic to a spined product of a left Clifford semigroup L and a right Clifford semigroup R with respect to a Clifford semigroup C and epimorphisms $\theta : L \to C$ and $\phi : R \to C$ such that for each $e \in E$ either $e\theta^{-1} = \{x\}$ for some $x \in E_{C_L}$ or $e\phi^{-1} = \{y\}$ for some $y \in E_{C_R}$.
(3) S is a completely regular subdirect product of a left inverse semigroup L and a right inverse semigroup R such that
$E_S \subseteq (E_{C_L} \times E_R) \cup (E_L \times E_{C_R})$
or, equivalently, $S \subseteq (C(L) \times R) \cup (L \times C(R)$ where E_{C_L} and E_{C_R} are the set of all idempotents in the centers $C(L)$ of L and $C(R)$ of R, respectively.

7 Some Problem

Definition 21 A regular semigroup is said to satisfy condition (A): if for any $a, t \in S$, $e \in E_S$, there exists $a_1 \in S$ such that $eat = a_1et$.

Clearly, all left Clifford semigroups are regular semigroups satisfying condition (A). For the converse, we consider the following examples:

Example 22 Let $S = \{a, b, c, d\}$ be a semigroup with the Cayley table:

	a	b	c	d
a	a	a	a	d
b	a	b	c	d
c	a	b	c	d
d	a	a	a	d

Here S is a band which is a regular semigroup satisfying condition (A). However, S is not a left Clifford semigroup since aS is not a subset of Sa, for any element a in S.

Also, it is easy to see that a right zero semigroup is a regular semigroup satisfying condition (A), but it is not left Clifford.

Theorem 23 *A regular semigroup S with condition (A) is completely regular.*

Problem. Describe this semigroup.

References

1. Clifford, A.H.: Semigroups admitting relative inverse. Ann. Math **42**, 1037–1049 (1942)
2. Clifford, A.H., Petrich, M.: Some classes of completely regular semigroups. J. Algebra **46**, 462–480 (1977)
3. Guo, Y.Q.: Structure of the weakly left C-semigroups. Chin. Sci. Bull. **41**, 462–467 (1996)
4. Guo, Y.Q., Shum, K.P., Sen, M.K.: LR-normal orthogroups. Sci. China. Ser. A Math. **49**(3), 330–341 (2006)
5. Petrich, M.: The structure of completely regular semigroups. Trans. Amer. Math. Soc. **189**, 211–236 (1974)
6. Petrich, M.: A structure theorem for completely regular semigroups. Proc. Amer. Math. Soc. **99**, 617–622 (1987)
7. Petrich, M., Reilly, N.R.: Completely Regular Semigroups. Wiley, New York (1999)
8. Howie, J.M.: An Introduction to Semigroup Theory. Academic Press, London (1976)
9. Kimura, N.: The structure of idempotent semigroups(1). Pacific J. Math. **8**, 257–275 (1958)
10. Sen, M.K., Ren, X.M., Shum, K.P.: A new structure theorem of LC- semigroups and a method for construction. Int. Math J. **3**(3), 283–295 (2003)
11. Sen, M.K., Ghosh, S., Pal, S.: On a class of subdirect products of left and right clifford semigroups. Commun. Algebra. **32**(7), 2609–2615 (2004)
12. Yamada, M.: Orthodox semigroups whose idempotents satisfy a certain identity. Semigroup Forum **6**, 113128 (1973)
13. Zhu, P.Y., Guo, Y.Q., Shum, K.P.: Structure and characteristics of left Clifford semigroups. Sci. China. Ser. A **35**(7), 791–805 (1992)

Certain Categories Derived from Normal Categories

A.R. Rajan

Abstract Normal categories are essentially the category of principal left(right) ideals of a regular semigroup which are used in describing the structure of regular semigroups. Several associated categories can be derived from normal categories which are of interest.

Keywords Normal categories · Ordered groupoid · Partial cones

Normal categories have been introduced by Nambooripad [7] to describe the structure of the categories of principal left [right] ideals of regular semigroups. These categories arise in the structure theory of regular semigroups. Several structure theories are available for regular semigroups and the differences arise in terms of the basic structures used in the theory. The structure theory in which the category of principal left ideals and the category of principal right ideals of the semigroup are the basic structures uses a relation between these categories known as cross connection. The cross-connections described for categories is a generalization of the cross-connection theory developed by Grillet [1] on partially ordered sets. The categories that arise in the theory of cross-connections are the normal categories. Here we consider some subcategories associated with a normal category and provide some of the relations among these subcategories.

1 Preliminaries

Here we introduce some notations and terminology on normal categories and related concepts. We follow generally the notations and terminology of Mac Lane [5] for concepts in category theory and Howie [2] for semigroup concepts.For normal categories the references are [4, 7, 8].

We consider a category $\mathcal{C}$ as consisting of a class of objects denoted by $v\mathcal{C}$ and for each pair of objects a, b a set $[a, b]_{\mathcal{C}}$ called the set of morphisms from a to b satisfying certain conditions. Here we write $f : a \to b$ to represent a morphism $f \in [a, b]_{\mathcal{C}}$. Also we may write $a = d(f)$ and $b = r(f)$ and call a the domain

A.R. Rajan (✉)
Department of Mathematics, University of Kerala, Thiruvananthapuram 695581, Kerala, India
e-mail: arrunivker@yahoo.com

P.G. Romeo et al. (eds.), *Semigroups, Algebras and Operator Theory*,
Springer Proceedings in Mathematics & Statistics 142,
DOI 10.1007/978-81-322-2488-4_5

of f and b the range or codomain of f. Further each of the sets $[a, b]_{\mathcal{C}}$ is called a homset. The following are the conditions.

1. For morphisms $f : a \to b$ and $g : c \to d$ a composition fg is defined only when $b = c$ and in this case the product fg is from a to d.
2. This composition is associative whenever defined, that is, if f, g, h are morphisms such that $(fg)h$ and $f(gh)$ are defined then $(fg)h = f(gh)$.
3. For each object $a \in v\mathcal{C}$ there is a morphism $1_a : a \to a$ such that $1_a f = f$ for all $f : a \to b$ and $g 1_a = g$ for all $g : b \to a$ for any $b \in v\mathcal{C}$.

The morphism 1_a is called the identity morphism at a. A morphism $f : a \to b$ is said to be an isomorphism or is said to be invertible if there exists a morphism $g : b \to a$ such that $fg = 1_a$ and $gf = 1_b$. A morphism $f : a \to b$ is said to be a monomorphism if it is cancellable on the right in compositions. That is, if $gf = hf$ for some morphisms $g, h : c \to a$ then $g = h$. A morphism f is said to be an epimorphism if it is cancellable on the left in compositions.

The concepts of functors and natural transformations are considered as in Mac Lane [5]. For categories $\mathcal{C}$ and $\mathcal{D}$, an isomorphism from $\mathcal{C}$ to $\mathcal{D}$ is a functor from $\mathcal{C}$ to $\mathcal{D}$ which is a bijection from $v\mathcal{C}$ to $v\mathcal{D}$ and a bijection on each homset.

A category $\mathcal{C}$ is said to be a small category when the class of objects $v\mathcal{C}$ is a set. We consider only small categories here.

The description of normal category follows the properties identified in the category of principal left ideals of a regular semigroup. We now describe this category. Let S be a regular semigroup. We denote by $\mathbb{L}(S)$ the category of principal left ideals of S whose objects and morphisms are given below. Objects are principal left ideals of S and morphisms are right translations on these ideals. A right translation on a semigroup S is mapping $\rho : S \to S$ such that $(xy)\rho = x(y\rho)$ for all $x, y \in S$. The restriction of a right translation on S to a subsemigroup is called a right translation on the subsemigroup.

Since S is a regular semigroup every principal left ideal is generated by an idempotent and so we have

$$v\mathbb{L}(S) = \{Se : e \in E(S)\}$$

where $E(S)$ is the set of all idempotents of S. A morphism $\rho : Se \to Sf$ in $\mathbb{L}(S)$ is a right translation on Se with image in Sf and so has the following description. Let $e\rho = u \in Sf$. Then since ρ is a right translation

$$u = e\rho = (ee)\rho = e(e\rho) = eu.$$

So we see that $u = euf \in eSf$ and for any $x \in Se$, $x\rho = (xe)\rho = x(e\rho) = xu$. Thus every morphism $\rho : Se \to Sf$ can be regarded as a right translation by an element $u \in eSf$ and we represent this morphism as $\rho(e, u, f)$. It may be observed that for $\rho(e, u, f) : Se \to Sf$ and $\rho(f, v, g) : Sf \to Sg$ the composition is given by $\rho(e, u, f)\rho(f, v, g) = \rho(e, uv, g)$.

The category $\mathbb{L}(S)$ has several additional features and we proceed to describe them.

A category $\mathcal{P}$ is said to be a preorder if for any objects a, b of $\mathcal{P}$ there is at most one morphism from a to b. If $[a, b]_{\mathcal{P}} \neq \emptyset$ we say $a \leq b$ and it can be seen that $\leq$ is a quasiorder(that is a reflexive and transitive relation) on $v\mathcal{P}$. If $[a, b] \cup [b, a]$ contains at most one element then $\leq$ is a partial order on $v\mathcal{P}$. In this case, $\mathcal{P}$ is said to be a strict preorder.

Definition 1.1 ([7]) A **category with subobjects** is a pair $(\mathcal{C}, \mathcal{P})$ where $\mathcal{C}$ is a small category and $\mathcal{P}$ is a subcategory of $\mathcal{C}$ satisfying the following,

1. $v\mathcal{C} = v\mathcal{P}$.
2. $\mathcal{P}$ is a strict preorder.
3. Every $f \in \mathcal{P}$ is a monomorphism in $\mathcal{C}$.
4. If $f, g \in \mathcal{P}$ and if $f = hg$ for some $h \in \mathcal{C}$ then $h \in \mathcal{P}$.

If $(\mathcal{C}, \mathcal{P})$ is a category with subobjects and if $f : a \to b$ is a morphism in $\mathcal{P}$ then we write $f = j(a, b)$ and is called the inclusion from a to b. Also we write $a \leq b$ in this case. Further, we often denote a category with subobjects as $\mathcal{C}$ in place of $(\mathcal{C}, \mathcal{P})$.

A cone in a category $(\mathcal{C}, \mathcal{P})$ with subobjects is a map γ from $v\mathcal{C}$ to the set of morphisms of $\mathcal{C}$ defined as follows.

Definition 1.2 A **cone** in a category with subobjects $\mathcal{C}$ is a map $\gamma : v\mathcal{C} \to \mathcal{C}$ satisfying the following.

1. There is an object c_γ called the vertex of the cone γ such that for each $a \in v\mathcal{C}$, $\gamma(a)$ is a morphism from a to c_γ. In this case $\gamma(a)$ is called the component of γ at a.
2. If $a, b \in v\mathcal{C}$ and $a \leq b$ then $\gamma(a) = j(a, b)\gamma(b)$.

A cone is said to be a **normal cone** if at least one component is an isomorphism. The vertex of a cone γ is usually denoted by c_γ.

Remark 1.3 The condition that at least one component of γ is an isomorphism is equivalent to the condition that at least one component is an epimorphism.

For, let c be the vertex of γ and let $\gamma(a) : a \to c$ be an epimorphism. So we can write $\gamma(a) = qu$ for a retraction q and an isomorphism u. Let $q : a \to a_0$ and $u : a_0 \to c$. Then $a_0 \leq a$ and $j(a_0, a)q = 1$ on a_0. Now from the property of cone, we get

$$\gamma(a_0) = j(a_0, a)\gamma(a) = j(a_0, a)qu = u.$$

Thus the component $\gamma(a_0)$ is an isomorphism.

Now we define normal category. The definition given below is essentially the definition for reductive normal category in the terminology of Nambooripad [7]. For convenience, we use the term normal category only.

Definition 1.4 A category with subobjects $(\mathcal{C}, \mathcal{P})$ is said to be a **normal category** if the following hold.

(N1) Every inclusion $j = j(a, b) : a \to b$ has a right inverse $q : b \to a$ in $\mathcal{C}$. Such a morphism q is called a retraction in $\mathcal{C}$.

(N2) Every morphism $f \in \mathcal{C}$ has a factorization in the form $f = quj$ where q is a retraction, u is an isomorphism, and j is an inclusion. Such a factorization is called a normal factorization.

(N3) For each $c \in v\mathcal{C}$ there exists a normal cone γ such that $\gamma(c) = 1_c$.

Remark 1.5 The definition of normal category may also be given with the condition (N3) above modified as follows.

(N3)': $(\mathcal{C}, \mathcal{P})$ contains normal cones. In this case $(\mathcal{C}, \mathcal{P})$ will give rise to another pair $(\mathcal{C}_0, \mathcal{P}_0)$ which satisfies the conditions (N1), (N2), and (N3). This can be seen as follows. Let $(\mathcal{C}, \mathcal{P})$ be a category with subobjects satisfying (N1), (N2) and (N3)'. Then we can produce a category with subobjects $(\mathcal{C}_0, \mathcal{P}_0)$ where $\mathcal{C}_0$ is a subcategory of $\mathcal{C}$ such that $(\mathcal{C}_0, \mathcal{P}_0)$ satisfies the conditions of Definition 2.1.

For this, consider the full subcategory $\mathcal{C}_0$ of $\mathcal{C}$ with

$$v\mathcal{C}_0 = \{c \in v\mathcal{C} : \text{ for some normal cone } \gamma \text{ in } \mathcal{C},\ \gamma(c) \text{ is an isomorphism}\}.$$

Let $\mathcal{P}_0$ be the full subcategory of $\mathcal{P}$ with $v\mathcal{P}_0 = v\mathcal{C}_0$. Then $(\mathcal{C}_0, \mathcal{P}_0)$ satisfies (N1), (N2), and (N3) of the above definition.

Example 1.6 It can be seen that the category $\mathbb{L}(S)$ of principal left ideals of a regular semigroup S is a normal category. Here for $Se \subseteq Sf$ the inclusion morphism is $j(Se, Sf) = \rho(e, e, f)$ and a retraction $q : Sf \to Se$ is $\rho(f, fe, e) = \rho(f, fe, fe)$. For an arbitrary morphism $\rho(e, u, f) : Se \to Sf$ a normal factorization is given by

$$\rho(e, u, f) = \rho(e, g, g)\rho(g, u, h)\rho(h, h, f)$$

for some $g \in E(R_u) \cap \omega(e)$ and $h \in E(L_u) \cap \omega(f)$.

Here R_u and L_u denotes the Green's equivalence classes and $\omega(e) = \{g \in E(S) : g \leq e$ in the natural partial order in $S\}$. It may further be noted that $E(R_u) \cap \omega(e)$ and $E(L_u) \cap \omega(f)$ are nonempty since S is regular.

The normal factorization of a morphism f given above is not in general unique. But if $f = quj$ and $f = q_1 u_1 j_1$ are normal factorizations then $qu = q_1 u_1$ and $j = j_1$ (cf. [7]). In this case $f^0 = qu$ is an epimorphism (that is left cancellative) and is called the epimorphic component of f. The epimorphic component f^0 is unique for f.

Each normal category $\mathcal{C}$ gives rise to a regular semigroup denoted by $T\mathcal{C}$, which is the semigroup of normal cones in $\mathcal{C}$ with composition defined as follows. For $\gamma, \delta \in T\mathcal{C}$, the product $\gamma * \delta$ is the normal cone given by

$$(\gamma * \delta)(a) = \gamma(a)(\delta(c_\gamma))^0.$$

It can be seen that $\gamma * \delta$ is a normal cone in this case and that $T\mathcal{C}$ is a regular semigroup with respect to the above product.

We often write $\gamma\delta$ in place of $\gamma * \delta$. Also if $f : c_\gamma \to c$ is a morphism, then we write $\gamma * f$ for the cone with components

$$(\gamma * f)(a) = \gamma(a)f.$$

It may noted that $\gamma * f$ is a normal cone whenever γ is normal and f is an epimorphism. The idempotents and Green's relations in $T\mathcal{C}$ are characterized as follows.

Proposition 1.7 ([7]) *Let $T\mathcal{C}$ be the semigroup of normal cones in a normal category $\mathcal{C} = (\mathcal{C}, \mathcal{P})$. Let γ, δ be normal cones in $\mathcal{C}$ with $c_\gamma = c$ and $c_\delta = d$. Then*

1. *γ is an idempotent in $T\mathcal{C}$ if and only if $\gamma(c) = 1_c$.*
2. *Let $\mathcal{L}$ be the Green's $\mathcal{L}$-relation in $T\mathcal{C}$. Then*

$$\gamma \; \mathcal{L} \; \delta \text{ if and only if } c = d.$$

The following theorem shows that every normal category arises as the category $\mathbb{L}(S)$ for a regular semigroup S.

Theorem 1.8 ([7]) *Let $\mathcal{C}$ be a normal category. Then $\mathcal{C}$ is isomorphic to $\mathbb{L}(S)$ where $S = T\mathcal{C}$.*

2 Subcategories of Retractions

Now we identify some subcategories in a normal category $\mathcal{C}$ and describe their relations with the semigroup $T\mathcal{C}$. The subcategory $\mathcal{P}$ of inclusions appear in the description of normal category. We can see that the collection of all retractions is a subcategory of $\mathcal{C}$ which can be described independent of inclusions.

Theorem 2.1 *Let $\mathcal{C} = (\mathcal{C}, \; \mathcal{P})$ be a normal category. Then*

1. *Let $q_1 : a \to b$ and $q_2 : b \to c$ be retractions in $\mathcal{C}$. Then $q_1q_2 : a \to c$ is a retraction in $\mathcal{C}$.*
2. *The set of all retractions of $\mathcal{C}$ forms a v-full subcategory of $\mathcal{C}$.*

Proof Since $q_1 : a \to b$ is a retraction, we have $b \leq a$ and $j(b, a)q_1 = 1_b$. Similarly $j(c, b)q_2 = 1_c$. Now $c \leq b \leq a$ and so $c \leq a$. Now

$$j(c, a)q_1q_2 = j(c, b)j(b, a)q_1q_2 = 1_c.$$

So q_1q_2 is a retraction.

Now to see that the subcategory of retractions is v-full, consider any $a \in v\mathcal{C}$. Then $1_a : a \to a$ is both a retraction and an inclusion. So the subcategory of retractions is v-full.

The following are the major properties of this subcategory.

Theorem 2.2 *Let $\mathcal{C} = (\mathcal{C}, \mathcal{P})$ be a normal category and $\mathcal{Q}$ be the subcategory of retractions in $\mathcal{C}$. Then the following hold for $a, b, c \in v\mathcal{C}$.*

1. *If $a \leq b$ then $\mathcal{Q}(b, a)$ is nonempty and $jq = 1_a$ for every $q \in \mathcal{Q}(b, a)$ and $j = j(a, b)$.*
2. *$\mathcal{Q}(b, a) = \emptyset$ if and only if $\mathcal{P}(a, b) = \emptyset$.*
3. *$\mathcal{P} \cap \mathcal{Q} = v\mathcal{C}$.*
4. *If $q_1, q_2 \in \mathcal{Q}$ and $q_1 = q_2 h$ for some $h \in \mathcal{C}$ with $r(h) \leq d(h)$ then $h \in \mathcal{Q}$.*

Proof The first three properties are clear from the definition of retraction. Now we prove the last statement. Let $q_1 : b \to a,\ q_2 : b \to c \in \mathcal{Q}$ and $h : c \to a \in \mathcal{C}$ with $a \leq c$. Now $j = j(a, c) \in \mathcal{P}$ and so there is a $q \in \mathcal{Q}$ such that $jq = 1_a$ such that $q_1 = q_2 h$. Let $j_1, j_2 \in \mathcal{P}$ be such that $j_1 q_1 = 1_a$ and $j_2 q_2 = 1_c$. Now $jj_2 = j_1$ and so

$$jh = j(j_2 q_2)h = (jj_2)(q_2 h) = j_1 q_1 = 1_a.$$

So $h \in \mathcal{Q}$.

Now we show that the conditions in the above theorem characterizes the subcategory of retractions in a normal category.

Theorem 2.3 *Let $\mathcal{C} = (\mathcal{C}, \mathcal{P})$ be a category with subobjects and $\mathcal{Q}$ be a v-full subcategory of $\mathcal{C}$. Then $\mathcal{Q}$ is the category of retractions associated with $(\mathcal{C}, \mathcal{P})$ if and only if $\mathcal{Q}$ satisfies the following.*

1. *For all $a \in v\mathcal{C}$, $\mathcal{Q}(a, a) = \{1_a\}$.*
2. *For $a, b \in v\mathcal{C}$ with $a \neq b$, if $\mathcal{Q}(b, a)$ is nonempty then $\mathcal{Q}(a, b)$ is empty and and there exists $j = j(a, b) \in \mathcal{P}$ such that $jq = 1_a$ for every $q \in \mathcal{Q}(b, a)$.*
3. *If $q_1, q_2 \in \mathcal{Q}$ and $q_1 = q_2 h$ for some $h \in \mathcal{C}$ with $r(h) \leq d(h)$ then $h \in \mathcal{Q}$.*

3 Groupoid of Isomorphisms

Another category we associate with a normal category is the groupoid of isomorphisms. It can be seen that this groupoid is an ordered groupoid. We start with the definition of ordered groupoid. A groupoid is a small category in which all morphisms are isomorphisms.

Definition 3.1 ([6]) A groupoid G with a partial order $\leq$ on G is said to be an ordered groupoid if the following hold.

(OG1) If $x, y, u, v \in G,\ xy, uv$ exist in G and if $x \leq u$ and $y \leq v$ then $xy \leq uv$.
(OG2) If $x \leq y$ then $x^{-1} \leq y^{-1}$.
(OG3) Let e be an identity in $G,\ x \in G$ and let e_x be the left identity of x. If $e \leq e_x$ then there exists a unique morphism $e * x \in G$ such that $e * x \leq x$ and the left identity of $e * x$ is e.

Remark 3.2 The element $e * x$ defined above is often called the restriction of x to e. Dually there is a concept of corestriction defined as follows. Let $x : a \to b$ in G and f be an identity in G with $f \le b$. Then $x * f = (f * x^{-1})^{-1}$ is called the corestriction of x to f. It may be noted that $x * f \le x$ and $r(x * f) = f$.

The next theorem describes the ordered groupoid associated with a normal category.

Theorem 3.3 *Let $\mathcal{C} = (\mathcal{C}, \mathcal{P})$ be a normal category. Then the set $G(\mathcal{C})$ of all isomorphisms in $\mathcal{C}$ is a v-full subcategory of $\mathcal{C}$ which is an ordered groupoid with partial order defined as follows. For $x : a \to b$ and $y : c \to d$ in $G(\mathcal{C})$,*

$$x \le y \text{ if } a \le c,\ b \le d \text{ and } xj(b, d) = j(a, c)y.$$

Proof Since the product of two isomorphisms is an isomorphism and identities are isomorphisms it follows that $G(\mathcal{C})$ is a v-full subcategory of $\mathcal{C}$. First observe that the partial order defined above coincides with the partial order of $v\mathcal{C}$ by considering elements of $v\mathcal{C}$ as identity morphisms in G.

Now let $x : a \to b$ be a morphism in $G = G(\mathcal{C})$. Then $d(x) = a$ and $r(x) = b$. Let $y : b \to c$ so that $xy : a \to c$. Let $u : a_1 \to b_1$ and $v : b_1 \to c_1$ so that $x \le u$ and $y \le v$. Then from the definition of $\le$ above, we see that

$$xyj(c, c_1) = xj(b, b_1)v = j(a, a_1)uv.$$

Therefore $xy \le uv$. Thus (OG1) holds. Similarly, we can see that (OG2) also holds.

To see (OG3) consider $x : a \to b$ in G and $e \le a$ in $v\mathcal{C}$. Now $j(e, a)x$ is a morphism in $\mathcal{C}$ and has a normal factorization. Let x_0 be the epimorphic part of this normal factorization which is unique. Then $j(e, a)x = x_0 j(r(x_0), b)$ and $x_0 \in G$. Further

$$x_0 = j(e, a)xq \tag{1}$$

where $q : b \to r(x_0)$ is a retraction and the product on the right side is independent of the choice of q. Define

$$e * x = x_0.$$

Then clearly $x_0 \le x$ and so (OG3) holds.

We now show that certain normal categories can be described in terms of the subcategories of inclusions, retractions, and the groupoid of isomorphisms. We consider the case where the category $\mathcal{P}$ of inclusions induces a semilattice order on the vertex set of $\mathcal{P}$. In this case for $a, b \in v\mathcal{P}$, we denote by $a \wedge b$ the meet of a and b in the semilattice.

Let $(G, \le)$ be an ordered groupoid with vertex set vG. Then the partially ordered set $(vG, \le)$ determines a preorder which is denoted by $\mathcal{P}(G)$. G is said to be semilattice ordered if $(vG, \le)$ is a semilattice. Let $\mathcal{Q}$ be the preorder which is dual to $\mathcal{P}$.

Then $v\mathcal{Q} = v\mathcal{P} = vG$ and for $a, b \in vG$, $\mathcal{Q}(a, b) \neq \emptyset$ if and only if $b \leq a$. We show that any $\mathcal{P}$, G, $\mathcal{Q}$ as above arises from a normal category.

Theorem 3.4 *Let $(G, \leq)$ be an ordered groupoid with semilattice order on identities. Let $G(\mathcal{P})$ be the preorder induced by the partial order $\leq$ on vG. Let $\mathcal{Q}$ be the dual of $\mathcal{P}$. Then $\mathcal{C} = \mathcal{Q} \otimes G \otimes \mathcal{P}$ is a normal category in which $\mathcal{P}$ is the subcategory of inclusions, $\mathcal{Q}$ is the subcategory of retractions, and G is the ordered groupoid of isomorphisms of $\mathcal{C}$ where*

$$\mathcal{Q} \otimes G \otimes \mathcal{P} = \{(q, u, j) \in \mathcal{Q} \times G \times \mathcal{P} : \ r(q) = d(u) \text{ and } r(u) = d(j)\}.$$

Proof First we describe a product of morphisms in $\mathcal{C} = \mathcal{Q} \otimes G \otimes \mathcal{P}$. Let (q, u, j), $(s, v, k) \in \mathcal{Q} \otimes G \otimes \mathcal{P}$ with $r(j) = d(s)$. Then the product is defined as

$$(q, u, j)(s, v, k) = (q', (u * h)(h * v), j')$$

where $h = r(u) \wedge d(v)$, $q' : d(q) \to d(u * h)$ and $j' : r(h * v) \to r(k)$. It follows from the definition of $u * h$ and $h * v$ that $d(u * h) \leq d(u)$ and $r(h * v) \leq r(v)$ so that $q' \in \mathcal{Q}$ and $j' \in \mathcal{P}$ exist. Clearly, the product is well defined and so $\mathcal{C}$ is a category with $v\mathcal{C} = vG$.

Let

$$\mathcal{P}' = \{(1_a, 1_a, j) : a \in vG \text{ and } j \in \mathcal{P} \text{ with } d(j) = a\}.$$

Then $\mathcal{P}'$ is the subcategory of inclusions of $\mathcal{C}$. Further for $(1_a, 1_a, j_1), (1_b, 1_b, j_2) \in \mathcal{P}'$ with $j_1 : a \to b$ it is easy to see that

$$(1_a, 1_a, j_1)(1_b, 1_b, j_2) = (1_a, 1_a, j_1 j_2).$$

Therefore, $\mathcal{P}'$ is isomorphic to $\mathcal{P}$. Similarly, we see that

$$\mathcal{Q}' = \{(q, 1_b, 1_b) : b \in vG \text{ and } q \in \mathcal{Q} \text{ with } r(q) = b\}$$

is the subcategory of retractions of $\mathcal{C}$ and $\mathcal{Q}'$ is isomorphic to $\mathcal{Q}$.

For any morphism $(q, u, j) \in \mathcal{C}$ with $q : a \to b$, $u : b \to c$ and $j : c \to d$ has a factorization

$$(q, u, j) = (q, 1_b, 1_b)(1_b, u, 1_c)(1_c, 1_c, j)$$

which is a normal factorization of (q, u, j).

Further for any $a \in v\mathcal{C}$ there is a normal cone σ with vertex a defined as follows. For any $b \in v\mathcal{C}$, $\sigma(b) = (q(b), 1_t, j(t, a))$ where $t = a \wedge b$ and $q(b) : b \to t$ is the unique morphism from b to t in $\mathcal{Q}$. Hence $\mathcal{C}$ is a normal category.

4 Groupoid of Partial Cones

Another category of interest associated with a normal category is the groupoid of partial cones. By a partial cone, we mean part of a normal cone where the base is suitably restricted. For any normal cone γ in a normal category $\mathcal{C}$ let

$$M(\gamma) = \{a \in v\mathcal{C} : \gamma(a) \text{ is an isomorphism}\}.$$

We denote by $m(\gamma)$, the partial cone corresponding to γ whose base is $M(\gamma)$ and components given by $m(\gamma)(a) = \gamma(a)$. Note that all the components of this partial cone are isomorphisms.

Let $P(M)$ denote the set of all partial cones in a normal category $\mathcal{C}$. We denote by $\mathcal{P}(M)$ the category whose vertex set is $P(M)$ and morphisms are as defined below. For $m(\gamma),\ m(\delta) \in P(M)$ a morphism from $m(\gamma)$ to $m(\delta)$ exists only when $M(\gamma) = M(\delta)$ and in this case $u = (\gamma(a))^{-1}\delta(a) : c_\gamma \to c_\delta$ for $a \in M(\gamma)$ is a morphism in $\mathcal{P}(M)$. Further $\mathcal{P}(M)(m(\gamma), m(\delta))$ is the set of all products of the form $uv : c_\gamma \to c_\delta$ where $u = (\gamma(a))^{-1}\sigma(a)$ and $v = (\sigma(b))^{-1}\delta(b)$ for some normal cone σ with $b \in M(\sigma) = M(\delta) = M(\gamma)$. Clearly $\mathcal{P}(M)$ becomes a category and all the morphisms are invertible. Thus $\mathcal{P}(M)$ is a groupoid. In fact $\mathcal{P}(M)$ is a subgroupoid of the groupoid $G(\mathcal{C})$ of isomorphisms of $\mathcal{C}$. So there is a partial order on $\mathcal{P}(M)$ induced by the partial order on $G(\mathcal{C})$.

Theorem 4.1 *Let $(\mathcal{C}, \mathcal{P})$ be a normal category. Then the groupoid $\mathcal{P}(M)$ of partial cones is an ordered groupoid.*

Proof First we define a partial order on $v\mathcal{P}(M)$ as follows. For normal cones γ and δ define

$$\gamma \leq \delta \text{ if } c_\gamma \leq c_\delta \text{ and } \gamma = \delta * q$$

for a retraction $q : c_\delta \to c_\gamma$. Now for $m(\gamma)$ and $m(\delta)$ in $v\mathcal{P}(M)$ define

$$m(\gamma) \leq m(\delta) \text{ if } \gamma \leq \delta.$$

Now let $\delta,\ \delta'$ be normal cones and $x : c_\delta \to c_{\delta'}$ be a morphism from $m(\delta)$ to $m(\delta')$ and let $m(\gamma) \leq m(\delta)$. Define

$$m(\gamma) * x = c_\gamma * x$$

where $c_\gamma * x$ is the restriction of x to c_γ defined in the ordered groupoid $G(\mathcal{C})$.

It remains to show that $c_\gamma * x$ defined above lies in $\mathcal{P}(M)$. By the definition of restriction in $G(\mathcal{C})$ and by Eq. (1), we see that

$$c_\gamma * x = j(c_\gamma, c_\delta)xq$$

where q is a retraction from $c_{\delta'}$ to $d' = r(c_\gamma * x)$.

First consider $x = (\delta(a))^{-1}\delta'(a)$ for some $a \in M(\delta)$. Then

$$c_\gamma * x = (\delta * q)(a')^{-1}\delta'(a')$$

where $a' = r(c_\gamma * (\delta(a))^{-1})$ and q is a retraction from c_δ to c_γ. So $c_\gamma * x \in \mathcal{P}(M)$. Similarly, we can see that for all $x \in \mathcal{P}(M)$ with $d(x) = c_\delta$ the restriction $c_\gamma * x \in \mathcal{P}(M)$. Thus $\mathcal{P}(M)$ is an ordered groupoid.

Remark 4.2 The ordered groupoid of partial cones can be used to describe the category of principal right ideals of the semigroup $T\mathcal{C}$ associated with the normal category $\mathcal{C}$.

References

1. Grillet, P.A.: Structure of regular semigroups, I. A representation; II Cross-connections; III The reduced case, Semigroup Forum **8**, 177–183, 254–265 (1974)
2. Howie, J.M.: Fundamentals of Semigroup Theory. Academic Press, New York (1995)
3. Lallement, G.: Semigroups and Combinatorial Applications. Wiley, New York (1979)
4. Lukose, S., Rajan, A.R.: Rings of normal cones. Indian J. Pure Appl. Math. **41**(5), 663–681 (2010)
5. Mac Lane, S.: Categories for the Working Mathematician. Springer, New York (1971)
6. Nambooripad, K.S.S.: Structure of regular semigroups I. Mem. Am. Math. Soc. **224** (1979)
7. Nambooripad, K.S.S.: Theory of Cross Connections, vol. 38. Centre for Mathematical Sciences, Trivandrum (1984)
8. Rajan, A.R.: Ordered groupoids and normal categories. In: Shum, K.P., et al. (eds.) Proceedings of International Conference on Semigroups and its Related Topics. Springer, New York (1995)

Semigroup Ideals and Permuting 3-Derivations in Prime Near Rings

Asma Ali, Clauss Haetinger, Phool Miyan and Farhat Ali

Abstract Let N be a near ring. A 3-additive map $\Delta : N \times N \times N \longrightarrow N$ is called a 3-derivation if the relations $\Delta(x_1x_2, y, z) = \Delta(x_1, y, z)x_2 + x_1\Delta(x_2, y, z)$, $\Delta(x, y_1y_2, z) = \Delta(x, y_1, z)y_2+y_1\Delta(x, y_2, z)$, and $\Delta(x, y, z_1z_2) = \Delta(x, y, z_1)z_2+z_1\Delta(x, y, z_2)$ are fulfilled, for all $x, y, z, x_i, y_i, z_i \in N$, $i = 1, 2$. The purpose of the present paper is to prove some commutativity theorems in the setting of a semigroup ideal of a 3-prime near ring admitting a permuting 3-derivation, thereby extending some known results of biderivations and permuting 3-derivations.

Keywords Near ring · 3-prime near rings · Semigroup ideals · Permuting 3-derivations

Mathematics Subject Classification (2010) 16N60 · 16W25 · 16Y30

1 Introduction

Throughout the paper, N denotes a zero-symmetric left near ring with multiplicative center Z, and for any pair of elements $x, y \in N$, $[x, y]$ denotes the commutator $xy - yx$ while the symbol (x, y) denotes the additive commutator $x + y - x - y$. A near ring N is called zero-symmetric if $0x = 0$, for all $x \in N$ (recall that left

A. Ali (✉) · P. Miyan · F. Ali
Department of Mathematics, Aligarh Muslim University, Aligarh 202002, India
e-mail: asma_ali2@rediffmail.com

P. Miyan
e-mail: phoolmiyan83@gmail.com

F. Ali
e-mail: 04farhatamu@gmail.com

C. Haetinger
Center of Exact and Technological Sciences, Univates University Centre,
Lajeado–RS 95900-000, Brazil
e-mail: chaet-@univates.br

P.G. Romeo et al. (eds.), *Semigroups, Algebras and Operator Theory*,
Springer Proceedings in Mathematics & Statistics 142,
DOI 10.1007/978-81-322-2488-4_6

distributivity yields that $x0 = 0$). The near ring N is said to be 3-prime if $xNy = \{0\}$ for $x, y \in N$ implies that $x = 0$ or $y = 0$. A near ring N is called n-torsion free, where n is a positive integer, if $(N, +)$ has no element of order n. For all $x \in N$, $C(x) = \{a \in N \mid ax = xa\}$ denotes the centralizer of x in N. A nonempty subset U of N is called a semigroup right (resp. semigroup left) ideal if $UN \subseteq U$ (resp. $NU \subseteq U$) and if U is both a semigroup right ideal and a semigroup left ideal, it is called a semigroup ideal. An additive map $d : N \longrightarrow N$ is called a derivation if the Leibniz rule $d(xy) = d(x)y + xd(y)$ holds for all $x, y \in N$. A biadditive map $D : N \times N \longrightarrow N$ (i.e., D is additive in both arguments) is said to be a biderivation if it satisfies the relations $D(xy, z) = D(x, z)y + xD(y, z)$ and $D(x, yz) = D(x, y)z + yD(x, z)$, for all $x, y, z \in N$.

A map $\Delta : N \times N \times N \longrightarrow N$ is said to be permuting if the equation $\Delta(x_1, x_2, x_3) = \Delta(x_{\pi(1)}, x_{\pi(2)}, x_{\pi(3)})$ holds for all $x_1, x_2, x_3 \in N$ and for every permutation π on $\{1, 2, 3\}$. A map $\delta : N \longrightarrow N$ defined by $\delta(x) = \Delta(x, x, x)$, for all $x \in N$ is called trace of Δ, where $\Delta : N \times N \times N \longrightarrow N$ is a permuting map. It is obvious that, in case $\Delta : N \times N \times N \longrightarrow N$ is a permuting map which is also 3-additive (i.e., additive in each argument), the trace of Δ satisfies the relation

$$\delta(x + y) = \delta(x) + 2\Delta(x, x, y) + \Delta(x, y, y) + \Delta(x, x, y) + 2\Delta(x, y, y) \\ + \delta(y) \text{ for all } x, y \in N.$$

Since we have

$$\Delta(0, y, z) = \Delta(0 + 0, y, z) = \Delta(0, y, z) + \Delta(0, y, z),$$

for all $y, z \in N$, we obtain $\Delta(0, y, z) = 0$, for all $y, z \in N$. Hence we get

$$0 = \Delta(0, y, z) = \Delta(x - x, y, z) = \Delta(x, y, z) + \Delta(-x, y, z)$$

and so we see that $\Delta(-x, y, z) = -\Delta(x, y, z)$, for all $x, y, z \in N$. Hence δ is an odd function.

A 3-additive map $\Delta : N \times N \times N \longrightarrow N$ is called a 3-derivation if the relations

$$\Delta(x_1x_2, y, z) = \Delta(x_1, y, z)x_2 + x_1\Delta(x_2, y, z),$$

$$\Delta(x, y_1y_2, z) = \Delta(x, y_1, z)y_2 + y_1\Delta(x, y_2, z),$$

and

$$\Delta(x, y, z_1z_2) = \Delta(x, y, z_1)z_2 + z_1\Delta(x, y, z_2)$$

are fulfilled for all $x, y, z, x_i, y_i, z_i \in N, i = 1, 2$. If Δ is permuting then, the above three relations are equivalent to each other.

For example, let N be a commutative near ring. A map $\Delta : N \times N \times N \longrightarrow N$ defined by $(x, y, z) \mapsto d(x)d(y)d(z)$, for all $x, y, z \in N$ is a permuting 3-derivation, where d is a derivation on N.

On the other hand, let S be a commutative near ring and let $N = \left\{ \begin{pmatrix} a & b \\ 0 & 0 \end{pmatrix} \mid a, b \in S \right\}$. It is clear that N is a noncommutative near ring under matrix addition and matrix multiplication. We define a map $\Delta : N \times N \times N \longrightarrow N$ by

$$\left(\begin{pmatrix} a_1 & b_1 \\ 0 & 0 \end{pmatrix}, \begin{pmatrix} a_2 & b_2 \\ 0 & 0 \end{pmatrix}, \begin{pmatrix} a_3 & b_3 \\ 0 & 0 \end{pmatrix} \right) \longmapsto \begin{pmatrix} 0 & a_1a_2a_3 \\ 0 & 0 \end{pmatrix}.$$

Then it is easy to see that Δ is a permuting 3-derivation.

In case of rings and near rings derivations and biderivations have received significant attention in recent years see [1, 2, 5, 6, 9–12]. Motivated by the notion of permuting 3-derivations in rings introduced by Ozturk in [4], Park and Jung [7] defined permuting 3-derivations in near rings and studied commutativity of a near ring admitting a 3-derivation satisfying certain conditions. In the present paper we obtain the results in the setting of a semigroup ideal of a 3-prime near ring admitting a permuting 3-derivation, thereby extending some known results on biderivations and 3-derivations.

2 Preliminary Results

We begin with the following lemmas.

Lemma 2.1 ([1, Lemma 1.2]) *Let N be a 3-prime near ring and Z be the center of N.*

(i) If $z \in Z\backslash\{0\}$, then z is not a zero divisor.
(ii) If $Z\backslash\{0\}$ contains an element z for which $z + z \in Z$, then $(N, +)$ is abelian.

Lemma 2.2 ([1, Lemma 1.3]) *Let N be a 3-prime near ring and U be a nonzero semigroup ideal of N.*

(i) If $x, y \in N$ and $xUy = \{0\}$, then $x = 0$ or $y = 0$.
(ii) If $x \in N$ and $xU = \{0\}$ or $Ux = \{0\}$, then $x = 0$.

Lemma 2.3 ([1, Lemma 1.5]) *If N is a 3-prime near ring and Z contains a nonzero semigroup left ideal or semigroup right ideal, then N is a commutative ring.*

Lemma 2.4 ([7, Lemma 2.4]) *Let N be a near ring and let $\Delta : N \times N \times N \longrightarrow N$ be a permuting 3-derivation. Then we have*

$$[\Delta(x, z, w)y + x\Delta(y, z, w)]v = \Delta(x, z, w)yv + x\Delta(y, z, w)v \quad \text{for all } v, w, x, y, z \in N.$$

Lemma 2.5 *Let N be a 3-prime near ring and U be a nonzero semigroup ideal of N. Let $\Delta : N \times N \times N \longrightarrow N$ be a nonzero 3-derivation. Then $\Delta(U, U, U) \neq \{0\}$.*

Proof Suppose that $\Delta(U, U, U) = \{0\}$. For any $u, v, w \in U$, we have

$$\Delta(u, v, w) = 0 \text{ for all } u, v, w \in U. \tag{2.1}$$

Substituting ux for u in (2.1), we get

$$\Delta(u, v, w)x + u\Delta(x, v, w) = 0 \text{ for all } u, v, w \in U \text{ and } x \in N.$$

Using (2.1), we get $U\Delta(x, v, w) = \{0\}$. Invoking Lemma 2.2 (*ii*), we have

$$\Delta(x, v, w) = 0 \text{ for all } v, w \in U \text{ and } x \in N. \tag{2.2}$$

Substituting vy for v in (2.2), we get

$$\Delta(x, v, w)y + v\Delta(x, y, w) = 0 \text{ for all } v, w \in U \text{ and } x, y \in N.$$

Using (2.2), we find $U\Delta(x, y, w) = \{0\}$ and Lemma 2.2 (*ii*), yields that

$$\Delta(x, y, w) = 0 \text{ for all } w \in U \text{ and } x, y \in N. \tag{2.3}$$

Substituting wz for w in (2.3), we obtain $U\Delta(x, y, z) = \{0\}$. Another appeal to Lemma 2.2 (*ii*), yields that $\Delta(x, y, z) = 0$, for all $x, y, z \in N$, which is a contradiction, since Δ is nonzero on N and hence theorem is proved.

Lemma 2.6 *Let N be a 3!-torsion-free near ring and U be a nonzero additive subgroup of N. Let $\Delta : N \times N \times N \longrightarrow N$ be a permuting 3-additive map with trace δ such that $\delta(x) = 0$, for all $x \in U$. Then we have $\Delta = 0$ on U.*

Proof For any $x, y \in U$, we have the relation

$$\delta(x + y) = \delta(x) + 2\Delta(x, x, y) + \Delta(x, y, y) + \Delta(x, x, y) + 2\Delta(x, y, y) + \delta(y)$$

and so, by the hypothesis, we get

$$2\Delta(x, x, y) + \Delta(x, y, y) + \Delta(x, x, y) + 2\Delta(x, y, y) = 0 \text{ for all } x, y \in U. \tag{2.4}$$

Substituting $-x$ for x in (2.4), we obtain

$$2\Delta(x, x, y) - \Delta(x, y, y) + \Delta(x, x, y) - 2\Delta(x, y, y) = 0 \text{ for all } x, y \in U. \tag{2.5}$$

On the other hand, for any $x, y \in U$,

$$\delta(y + x) = \delta(y) + 2\Delta(y, y, x) + \Delta(y, x, x) + \Delta(y, y, x) + 2\Delta(y, x, x) + \delta(x)$$

and thus, by the hypothesis and using the fact that Δ is permuting, we have

$$2\Delta(x, y, y) + \Delta(x, x, y) + \Delta(x, y, y) + 2\Delta(x, x, y) = 0 \text{ for all } x, y \in U. \quad (2.6)$$

Comparing (2.4) and (2.5), we get

$$2\Delta(x, y, y) + \Delta(x, x, y) + \Delta(x, y, y) = \Delta(x, x, y) - 3\Delta(x, y, y)$$

which implies that

$$2\Delta(x, y, y) + \Delta(x, x, y) + \Delta(x, y, y) + 2\Delta(x, x, y) = \Delta(x, x, y) - 3\Delta(x, y, y) + 2\Delta(x, x, y).$$

Hence it follows from (2.6) that

$$\Delta(x, x, y) - 3\Delta(x, y, y) + 2\Delta(x, x, y) = 0 \text{ for all } x, y \in U. \quad (2.7)$$

Substituting $-x$ for x in (2.7), we find

$$\Delta(x, x, y) + 3\Delta(x, y, y) + 2\Delta(x, x, y) = 0 \text{ for all } x, y \in U. \quad (2.8)$$

Comparing (2.7) and (2.8), we obtain

$$6\Delta(x, y, y) = 0 \text{ for all } x, y \in U.$$

Since N is 3!-torsion free, we get

$$\Delta(x, y, y) = 0 \text{ for all } x, y \in U. \quad (2.9)$$

Substituting $y + z$ for y in (2.9) and linearizing (2.9) we obtain

$$\Delta(x, y, z) = 0 \text{ for all } x, y \in U,$$

i.e., $\Delta = 0$ on U which completes the proof.

Lemma 2.7 *Let N be a 3-prime near ring and U be a nonzero semigroup ideal of N. Let $\Delta : N \times N \times N \longrightarrow N$ be a 3-derivation. If $x \in N$ and $\Delta(U, U, U)x = \{0\}$ (or $x\Delta(U, U, U) = \{0\}$), then either $x = 0$ or $\Delta = 0$ on U.*

Proof Let

$$\Delta(y, z, w)x = 0 \text{ for all } y, z, w \in U.$$

Substituting $y = vy$ we get $\Delta(v, z, w)yx = 0$, for all $y, z, v, w \in U$, i.e., $\Delta(v, z, w)Ux = \{0\}$, for all $v, z, w \in U$. Invoking Lemma 2.2 (i) either $x = 0$ or $\Delta = 0$ on U.

Lemma 2.8 *Let N be a 3!-torsion-free 3-prime near ring and U be a nonzero additive subgroup and a semigroup ideal of N. Let $\Delta : N \times N \times N \longrightarrow N$ be permuting 3-derivation with trace δ and $x \in N$ such that $x\delta(y) = 0$, for all $y \in U$. Then either $x = 0$ or $\Delta = 0$ on U.*

Proof For any $y, z \in U$, we have

$$\delta(y + z) = \delta(y) + 2\Delta(y, y, z) + \Delta(y, z, z) + \Delta(y, y, z) + 2\Delta(y, z, z) + \delta(z).$$

By hypothesis

$$2x\Delta(y, y, z) + x\Delta(y, z, z) + x\Delta(y, y, z) + 2x\Delta(y, z, z) = 0 \text{ for all } y, z \in U. \tag{2.10}$$

Substituting $-y$ for y in (2.10), it follows that

$$2x\Delta(y, y, z) - x\Delta(y, z, z) + x\Delta(y, y, z) - 2x\Delta(y, z, z) = 0 \text{ for all } y, z \in U. \tag{2.11}$$

On the other hand,

$$\delta(z + y) = \delta(z) + 2\Delta(z, z, y) + \Delta(z, y, y) + \Delta(z, z, y) + 2\Delta(z, y, y) + \delta(y)$$

Again using hypothesis, we have

$$2x\Delta(z, z, y) + x\Delta(z, y, y) + x\Delta(z, z, y) + 2x\Delta(z, y, y) = 0.$$

Since Δ is permuting, we get

$$2x\Delta(y, z, z) + x\Delta(y, y, z) + x\Delta(y, z, z) + 2x\Delta(y, y, z) = 0 \text{ for all } y, z \in U. \tag{2.12}$$

Comparing (2.10) and (2.11), we get

$$2x\Delta(y, z, z) + x\Delta(y, y, z) + x\Delta(y, z, z) = x\Delta(y, y, z) - 3x\Delta(y, z, z),$$

i.e.,

$$\begin{aligned} 2x\Delta(y, z, z) + x\Delta(y, y, z) + x\Delta(y, z, z) + 2x\Delta(y, y, z) = x\Delta(y, y, z) \\ - 3x\Delta(y, z, z) + 2x\Delta(y, y, z). \end{aligned}$$

Now, from (2.12), we obtain

$$x\Delta(y, y, z) - 3x\Delta(y, z, z) + 2x\Delta(y, y, z) = 0 \text{ for all } y, z \in U. \tag{2.13}$$

Substituting $-y$ for y in (2.13), we find

$$x\Delta(y, y, z) + 3x\Delta(y, z, z) + 2x\Delta(y, y, z) = 0 \text{ for all } y, z \in U. \tag{2.14}$$

Comparing (2.13) and (2.14), we obtain

$$6x\Delta(y, z, z) = 0 \text{ for all } y, z \in U.$$

Since N is 3!-torsion free, we get

$$x\Delta(y, z, z) = 0 \text{ for all } y, z \in U. \tag{2.15}$$

Substituting $z + w$ for z in (2.15), we have

$$x\Delta(w, y, z) = 0 \text{ for all } w, y, z \in U. \tag{2.16}$$

Hence by Lemma 2.7 either $x = 0$ or $\Delta = 0$ on U.

3 Main Results

Theorem 3.1 *Let N be a 3-prime near ring and U be a nonzero semigroup ideal of N which is closed under addition. Let $\Delta : N \times N \times N \longrightarrow N$ be a nonzero 3-derivation such that $\Delta(U, U, U) \subseteq Z$. Then $(N, +)$ is abelian.*

Proof By hypothesis $\Delta(x, y, z) \in Z$, for all $x, y, z \in U$. Since Δ is nonzero on U by Lemma 2.5, there exists nonzero elements $x_0, y_0, z_0 \in U$ such that $0 \neq \Delta(x_0, y_0, z_0) \in Z \setminus \{0\}$ and $\Delta(x_0, y_0, z_0 + z_0) = \Delta(x_0, y_0, z_0) + \Delta(x_0, y_0, z_0) \in Z$. Hence $(N, +)$ is abelian by Lemma 2.1 (*ii*).

Theorem 3.2 *Let N be a 3!-torsion-free 3-prime near ring and U be a nonzero additive subgroup and a semigroup ideal of N. Let $\Delta : N \times N \times N \longrightarrow N$ be a nonzero permuting 3-derivation with trace δ such that $\Delta(U, U, U) \subseteq Z$ and $\delta(U) \subseteq U$. Then N is a commutative ring.*

Proof By hypothesis

$$w\Delta(x, y, z) = \Delta(x, y, z)w \text{ for all } x, y, z \in U \text{ and } w \in N. \tag{3.1}$$

Replacing x by xv in (3.1) and using Lemma 2.4, we get

$$w\Delta(x, y, z)v + wx\Delta(v, y, z) = \Delta(x, y, z)vw + x\Delta(v, y, z)w$$
$$\text{for all } v, x, y, z \in U \text{ and } w \in N.$$

Using hypothesis and Theorem 3.1, we have

$$\Delta(x, y, z)[w, v] = \Delta(v, y, z)[x, w] \text{ for all } v, x, y, z \in U \text{ and } w \in N. \tag{3.2}$$

Replacing x by $\delta(u)$ in (3.2), we obtain

$$\Delta(\delta(u), y, z)[w, v] = 0 \text{ for all } u, v, y, z \in U \text{ and } w \in N. \tag{3.3}$$

Assume that

$$\Delta(\delta(u), y, z) = 0.$$

Substituting $u + x$ for u and using the hypothesis, we obtain

$$\Delta(\Delta(u, u, x), y, z) + \Delta(\Delta(u, x, x), y, z) = 0 \text{ for all } u, x, y, z \in U. \tag{3.4}$$

Replacing u by $-u$ in (3.4) and using (3.4), we obtain

$$\Delta(\Delta(u, u, x), y, z) = 0 \text{ for all } u, x, y, z \in U. \tag{3.5}$$

Now replacing x by ux in (3.5) and using (3.3), we find

$$\delta(u)\Delta(x, y, z) + \Delta(u, y, z)\Delta(u, u, x) = 0,$$

and by hypothesis

$$\delta(u)\Delta(x, y, z) + \Delta(u, u, x)\Delta(u, y, z) = 0 \text{ for all } u, x, y, z \in U. \tag{3.6}$$

Taking $u = y = x$ in (3.6), we obtain

$$\delta(x)\Delta(x, x, z) = 0 \text{ for all } x, z \in U. \tag{3.7}$$

If $\Delta(x, x, z)$ is nonzero element of Z, then $\delta(x) = 0$, for all $x \in U$ by Lemma 2.1 (i). On the other hand if $\Delta(x, x, z) = 0$, for all $x, z \in U$, we have $\Delta(x, x, x) = 0$, i.e., $\delta(x) = 0$. Hence in both the cases $\delta(x) = 0$, for all $x \in U$ and using Lemma 2.6 we have $\Delta = 0$ on U—a contradiction by Lemma 2.5. Thus $\Delta(\delta(u), y, z)$ is a nonzero element of Z and by Lemma 2.1 (i), (3.3) yields that $U \subseteq Z$. Hence N is a commutative ring by Lemma 2.3.

Theorem 3.3 *Let N be a 3!-torsion-free 3-prime near ring and U be a nonzero additive subgroup and a semigroup ideal of N. Let $\Delta : N \times N \times N \longrightarrow N$ be a nonzero permuting 3-derivation with trace δ such that $\delta(x), \delta(x) + \delta(x) \in C(\Delta(U, U, U))$, for all $x \in U$. Then $(N, +)$ is abelian.*

Proof For all $t, u, v, x, w \in U$

$$\begin{aligned}
\Delta(u+t, v, w)(\delta(x)+\delta(x)) \\
&= (\delta(x)+\delta(x))\Delta(u+t, v, w) \\
&= (\delta(x)+\delta(x))[\Delta(u, v, w)+\Delta(t, v, w)] \\
&= (\delta(x)+\delta(x))\Delta(u, v, w)+(\delta(x)+\delta(x))\Delta(t, v, w) \\
&= \Delta(u, v, w)(\delta(x)+\delta(x))+\Delta(t, v, w)(\delta(x)+\delta(x)) \\
&= \Delta(u, v, w)\delta(x)+\Delta(u, v, w)\delta(x)+\Delta(t, v, w)\delta(x) \\
&\quad + \Delta(t, v, w)\delta(x) \\
&= \delta(x)\Delta(u, v, w)+\delta(x)\Delta(u, v, w)+\delta(x)\Delta(t, v, w) \\
&\quad + \delta(x)\Delta(t, v, w) \\
&\quad \text{for all } t, u, v, w, x \in U.
\end{aligned} \tag{3.8}$$

and

$$\begin{aligned}
\Delta(u+t, v, w)(\delta(x)+\delta(x)) \\
&= \Delta(u+t, v, w)\delta(x)+\Delta(u+t, v, w)\delta(x) \\
&= \delta(x)\Delta(u+t, v, w)+\delta(x)\Delta(u+t, v, w) \\
&= \delta(x)[\Delta(u, v, w)+\Delta(t, v, w)] \\
&\quad + \delta(x)[\Delta(u, v, w)+\Delta(t, v, w)] \\
&= \delta(x)\Delta(u, v, w)+\delta(x)\Delta(t, v, w)+\delta(x)\Delta(u, v, w) \\
&\quad + \delta(x)\Delta(t, v, w) \\
&\quad \text{for all } t, u, v, w, x \in U.
\end{aligned} \tag{3.9}$$

Comparing (3.8) and (3.9), we obtain

$$\delta(x)\Delta((u, t), v, w) = 0 \text{ for all } t, u, v, w, x \in U.$$

By hypothesis we get

$$\Delta((u, t), v, w)\delta(x) = 0 \text{ for all } t, u, v, w, x \in U.$$

Hence it follows from Lemma 2.8, that

$$\Delta((u,t),v,w)=0 \text{ for all } t,u,v,w\in U. \tag{3.10}$$

Substituting uz for u and ut for t in (3.10) to get

$$0=\Delta(u(z,t),v,w)=\Delta(u,v,w)(z,t)+u\Delta((z,t),v,w),$$

i.e.,

$$\Delta(u,v,w)(z,t)=0 \text{ for all } t,u,v,w,z\in U. \tag{3.11}$$

By Lemma 2.7 either $(z,t)=0$ or $\Delta=0$ on U. Later yields a contradiction by Lemma 2.5. Hence $(z,t)=0$, for all $z,t\in U$. Substituting zr for z and zs for t, for all $r,s\in N$, we have

$$z(r,s)=0 \text{ for all } z\in U \text{ and } r,s\in N,$$

i.e.,

$$U(r,s)=\{0\} \text{ for all } r,s\in N. \tag{3.12}$$

Invoking Lemma 2.2 (*ii*), $(r,s)=0$, for all $r,s\in N$ and $(N,+)$ is abelian.

Theorem 3.4 *Let N be a 3!-torsion-free 3-prime near ring and U be a nonzero additive subgroup and a semigroup ideal of N. Let $\Delta: N\times N\times N\longrightarrow N$ be a nonzero permuting 3-derivation with trace δ such that $\delta(x),\delta(x)+\delta(x)\in C(\Delta(U,U,U))$, for all $x\in U$ and $\delta(U)\subseteq U$. Then N is a commutative ring.*

Proof By the hypothesis for all $u,v,w,x\in U$

$$[\delta(x),\Delta(u,v,w)]=0. \tag{3.13}$$

Replacing x by $x+y$ in (3.13) and using Theorem 3.3, we obtain

$$[\Delta(x,x,y),\Delta(u,v,w)]+[\Delta(x,y,y),\Delta(u,v,w)]=0 \text{ for all } u,v,w,x,y\in U. \tag{3.14}$$

Setting $y=-y$ in (3.14) and comparing the result with (3.14), we obtain

$$[\Delta(x,y,y),\Delta(u,v,w)]=0 \text{ for all } u,v,w,x,y\in U. \tag{3.15}$$

Replacing y by $y+z$ in (3.15), using (3.15) and the fact that Δ is permuting, we have

$$[\Delta(x,y,z),\Delta(u,v,w)]=0 \text{ for all } u,v,w,x,y,z\in U.$$

$$\Delta(x, y, z)\Delta(u, v, w) = \Delta(u, v, w)\Delta(x, y, z) \text{ for all } u, v, w, x, y, z \in U. \quad (3.16)$$

Substituting ut for u in (3.16), we find

$$\Delta(u, v, w)t\Delta(x, y, z) - \Delta(x, y, z)\Delta(u, v, w)t + u\Delta(t, v, w)\Delta(x, y, z)$$
$$- \Delta(x, y, z)u\Delta(t, v, w) = 0 \text{ for all } t, u, v, w, x, y, z \in U. \quad (3.17)$$

Substituting $\delta(u)$ for u in (3.17) and using (3.16), we get

$$\Delta(\delta(u), v, w)[t, \Delta(x, y, z)] = 0 \text{ for all } t, u, v, w, x, y, z \in U. \quad (3.18)$$

Replacing w by ws in (3.18), we have

$$\Delta(\delta(u), v, w)s[t, \Delta(x, y, z)] = 0 \text{ for all } s, t, u, v, w, x, y, z \in U,$$

i.e.,

$$(\delta(u), v, w)U[t, \Delta(x, y, z)] = 0 \text{ for all } t, u, v, w, x, y, z \in U.$$

By Lemma 2.2 (i), either $\Delta(\delta(u), v, w) = 0$ or $[t, \Delta(x, y, z)] = 0$.

Assume that

$$\Delta(\delta(u), v, w) = 0 \text{ for all } u, v, w \in U. \quad (3.19)$$

Substituting $u + x$ for u in (3.19) and using the hypothesis, we obtain

$$\Delta(\Delta(u, u, x), v, w) + \Delta(\Delta(u, x, x), v, w) = 0 \text{ for all } u, x, v, w \in U. \quad (3.20)$$

Setting $u = -u$ in (3.20) and comparing the result with (3.20), we find that

$$\Delta(\Delta(u, u, x), v, w) = 0 \text{ for all } u, x, v, w \in U. \quad (3.21)$$

Substituting ux for x in (3.21) and applying (3.19), we obtain

$$\delta(u)\Delta(x, v, w) + \Delta(u, v, w)\Delta(u, u, x) = 0$$

and so it follows from the hypothesis that

$$\delta(u)\Delta(x, v, w) + \Delta(u, u, x)\Delta(u, v, w) = 0 \text{ for all } u, x, v, w \in U. \quad (3.22)$$

Taking $u = v = x$ in (3.22) and using hypothesis, we obtain

$$\Delta(x, x, w)\delta(x) = 0 \text{ for all } x, w \in U. \quad (3.23)$$

By Lemma 2.8, $\Delta(x, x, w) = 0$. Replacing w by x, we have $\delta(x) = 0$, for all $x \in U$ and Lemma 2.6 yields that $\Delta = 0$ on U—a contradiction by Lemma 2.5.

Consequently, we arrive at

$$[t, \Delta(x, y, z)] = 0 \text{ for all } t, x, y, z \in U.$$

Substituting rt for t, we get

$$[r, \Delta(x, y, z)]t = 0 \text{ for all } t, x, y, z \in U \text{ and } r \in N,$$

i.e.,

$$[r, \Delta(x, y, z)]U = \{0\} \text{ for all } x, y, z \in U \text{ and } r \in N.$$

By Lemma 2.2 (*ii*), we get

$$[r, \Delta(x, y, z)] = 0 \text{ for all } x, y, z \in U \text{ and } r \in N,$$

i.e., $\Delta(x, y, z) \in Z$, for all $x, y, z \in U$. Hence N is a commutative ring by Theorem 3.2.

The following example justifies that Theorem 3.4 does not hold for an arbitrary near ring and conditions U to be a semigroup ideal of N and $\delta(U) \subseteq U$ in the hypothesis are essential.

Example 3.1 Let S be any commutative near ring. Let $N = \left\{ \begin{pmatrix} x & y \\ 0 & 0 \end{pmatrix} \mid x, y \in S \right\}$ and $U = \left\{ \begin{pmatrix} x & 0 \\ 0 & 0 \end{pmatrix} \mid x \in S \right\}$. Then N is a near ring and U is an additive subgroup of N but not a semigroup ideal of N. Define $\Delta : N \times N \times N \longrightarrow N$ by

$$\left(\begin{pmatrix} x_1 & y_1 \\ 0 & 0 \end{pmatrix}, \begin{pmatrix} x_2 & y_2 \\ 0 & 0 \end{pmatrix}, \begin{pmatrix} x_3 & y_3 \\ 0 & 0 \end{pmatrix} \right) \longmapsto \begin{pmatrix} 0 & x_1x_2x_3 \\ 0 & 0 \end{pmatrix} \text{ and } \delta : N \longrightarrow N$$

such that $\delta(y) = \Delta(y, y, y)$, for all $y \in N$. Then Δ is a permuting 3-derivation on N with trace δ. It can be easily verified that $\delta(U) \nsubseteq U$ and $\delta(x), \delta(x) + \delta(x) \in C(\Delta(u, v, w))$, for all $u, v, w, x \in U$. Neither $(N, +)$ is abelian nor N under multiplication is commutative.

References

1. Bell, H.E.: On derivations in near rings II. Math. Appl. Dordr. **426**, 191–197 (1997). (Kluwer Academic Publishers)
2. Ceven, Y., Ozturk, M.A.: Some properties of symmetric *bi*-(σ, τ)-derivations in near-rings. Commun. Korean Math. Soc. **22**(4), 487–491 (2007)

3. Jung, Y.S., Park, K.H.: On prime and semiprime rings with permuting 3-derivations. Bull. Korean Math. Soc. **44**(4), 789–794 (2007)
4. Ozturk, M.A.: Permuting tri-derivations in prime and semiprime rings. East Asian Math. J. **15**(2), 177–190 (1999)
5. Ozturk, M.A., Yazarli, H.: Some results on symmetric bi-(σ, τ)-derivations in near rings. Miskolc Math. Notes **11**(2), 169–173 (2010)
6. Ozturk, M.A., Jun, Y.B.: On the trace of symmetric bi-derivations in near rings. Int. J. Pure Appl. Math. **17**(1), 93–100 (2004)
7. Park, K.H., Jung, Y.S.: On permuting 3-derivations and commutativity in prime near rings. Commun. Korean Math. Soc. **25**(1), 1–9 (2010). doi:10.4134/CKMS.2010.25.1.001
8. Pilz, G.: Near-rings, 2nd edn. North Holland, Amsterdam (1983)
9. Posner, E.C.: Derivations in prime rings. Proc. Am. Math. Soc. **8**, 1093–1100 (1957)
10. Uckum, M., Ozturk, M.A.: On trace of symmetric bi-gamma derivations in gamma near rings. Houst. J. Math. **33**(2), 223–339 (2007)
11. Vukman, J.: Two results concerning symmetric biderivations on prime rings. Aequ. Math. **40**, 181–189 (1990)
12. Vukman, J.: Symmetric biderivations on prime and semiprime rings. Aequ. Math. **38**, 245–254 (1989)

Biordered Sets and Regular Rings

P.G. Romeo

Abstract Biordered sets were introduced in [3] to describe the structure of regular semigroups. In [1] it is shown that the ideals of a regular ring forms a complemented modular lattices. Here we describe the biordered set of such a regular ring.

Keywords Biordered sets · Regular ring · Modular lattice

Mathematics Subject Classification (1991) 20M10

1 Preliminaries

First, we recall the basic definitions and results regarding lattices and rings needed in the sequel. Let L be a class of elements $a, b, \ldots$ together with a binary relation $<$ between pairs of elements of L satisfying $a \vee b$ and $a \wedge b$ exists, for all $a, b \in L$, then L is called a lattice. Further if $\vee S$ and $\wedge S$ exist for all $S \subseteq L$, then L is called a complete lattice.

Definition 1 A complemented lattice is a bounded lattice (with least element 0 and greatest element 1) in which every element a has a complement b such that

$$a \vee b = 1 \quad \text{and} \quad a \wedge b = 0.$$

Definition 2 Let L be a lattice

(1) L is said to be distributive if it satisfies the distributive laws

$$a \wedge (b \vee c) = (a \wedge b) \vee (a \wedge c), \text{ and}$$
$$a \vee (b \wedge c) = (a \vee b) \wedge (a \vee c), \ \forall a, b, c \in L$$

P.G. Romeo (✉)
Department of Mathematics, Cochin University of Science and Technology, Kochi, Kerala, India
e-mail: romeo_parackal@yahoo.com

P.G. Romeo et al. (eds.), *Semigroups, Algebras and Operator Theory*, Springer Proceedings in Mathematics & Statistics 142,
DOI 10.1007/978-81-322-2488-4_7

(2) L is said to be modular if it satisfies the modular law

$$a \geq c \Rightarrow a \wedge (b \vee c) = (a \wedge b) \vee c, \ \forall a, b, c \in L$$

A complemented distributive lattice is called a *Boolean algebra*. Obviously, a Boolean algebra is a complemented modular lattice. A typical example of a complemented modular lattice which is not a Boolean algebra is the lattice of linear manifolds of a finite dimensional vector space.

A set S together with an associative binary operation is called a semigroup. An element a in a semigroup S is called regular if there exists an element $a' \in S$ such that $aa'a = a$, if every element of S is regular then S is a regular semigroup. An element $e \in S$ such that $e \cdot e = e$ is called an idempotent and the set of all idempotents in S will be denoted by $E(S)$.

Let S be a semigroup for $a \in S$ the smallest left ideal containing a is $S^1a = Sa \cup a$ and is called the principal left ideal generated by a. An equivqlence relation $\mathcal{L}$ on S is defined by $a\mathcal{L}b$ if and only if $S^1a = S^1b$. Similarly, we can define the relation $\mathcal{R}$ by $a\mathcal{R}b$ if and only if $aS^1 = bS^1$. The intersection of $\mathcal{L}$ and $\mathcal{R}$ is the relation $\mathcal{H}$ and their union $\mathcal{D}$. It is easy to observe that these are equivalence relations and these relations are termed as Green's equivalences.

Lemma 1 *A semigroup S is regular if each $\mathcal{L}$ class and each $\mathcal{R}$ class of S contain idempotents.*

Lemma 2 *A semigroup S is an inverse semigroup if each $\mathcal{L}$ class and each $\mathcal{R}$ class of S contain a unique idempotent.*

Remark 1 If a is a regular element in a semigroup S then there exists a' such that $aa'a = a$, hence we have both aa' and $a'a$ which are idempotents and so the above lemmas follows.

Definition 3 Let S be a semigroup with 0 and X any subset of S, then

$$X^r = \{y \in S : xy = 0 \text{ for all } x \in X\}$$

is the right annihilator of X. If $X = \{x\}$ then x^r is the right annihilator of x.

Similarly, one can define the left annihilator X^l and x^l.

A Baer semigroup is a multiplicative semigroup with zero in which the left [right] annihilator of each element x is a principal left [right] ideal.

Definition 4 A regular semigroup S for which the set of the principal left [right] ideals coincides with the set of left [right] annihilators of elements of S is called a strongly regular Baer semigroup.

Let P be a partially ordered set and $\Phi : P \to P$ be an isotone mapping, then Φ is called normal

(1) $im\Phi$ is a principal ideal of P
(2) whenever $x\Phi = y$, then there exists some $z \leq x$ such that Φ maps the principal ideal $P(z)$ isomorphically onto the principal ideal $P(y)$.

Definition 5 The partially ordered set P is called regular if for every $e \in P,\ P(e) = im\Phi$ for some normal mapping $\Phi : P \to P$ with $\Phi^2 = \Phi$

If P is a regular partially ordered set , then the set $S(P)$ $[S^*(P)]$ of normal mappings of P into itself, considered as left [right] operators form a regular semigroup with the composition of maps as binary operation and $S(P)/\mathcal{R} = S^*(P)/\mathcal{L} = P$. The following is an interesting lemma in this context.

Lemma 3 *Let S be a regular semigroup. Then $S/\mathcal{L}$ and $S/\mathcal{R}$ are regular partially ordered sets. For any $y \in S$ the mappings*

$$r_y : S/\mathcal{R} \to S/\mathcal{R},\ R_x \to r_y R_x = R_{xy}$$

and

$$l_y : S/\mathcal{L} \to S/\mathcal{L},\ L_x \to L_x l_y = L_{xy}$$

are normal mappings.

Definition 6 A ring $\mathcal{R}$ is a set of elements $x, y, \ldots$ together with two binary operations ' $+$ ' and ' $\cdot$ ' with the following properties:

(1) $(\mathcal{R}, +)$ is an abelian group.
(2) $(\mathcal{R}, \cdot)$ is a semigroup.
(3) $x \cdot (y + z) = x \cdot y + x \cdot z$ and $(y + z) \cdot x = y \cdot x + z \cdot x$, for all $x, y, z \in R$.

Further if there exists an element 1 with the property that $x \cdot 1 = 1 \cdot x = x$ for every $x \in \mathcal{R}$, then the ring is called a ring with unity. In the following our ring $\mathcal{R}$ is always regarded as a ring with unity. An element $e \in \mathcal{R}$ is said to be an idempotent if $e^2 = e$. Obviously, 1 is an idempotent element.

Definition 7 A ring $\mathcal{R}$ is called a division algebra or a field if for every $x \neq 0$ there exists a y such that $x \cdot y = y \cdot x = 1$. This y is unique and will be denoted by x^{-1}.

1.1 Ideals and Modular Lattices

A subset a of a ring $\mathcal{R}$ is called right ideal in case

$$x + y \in a,\ \ xz \in a$$

for each $x, y \in a$ and $z \in \mathcal{R}$. Similarly, we can define the left ideal in $\mathcal{R}$. We call a an ideal if it is both a right and a left ideal. The set of all right (left) ideals of $\mathcal{R}$ is denoted by $R_{\mathcal{R}}(L_{\mathcal{R}})$.

It is easy to note that the intersection of any class of right(left) ideals is again a right (left) ideal and also for any $a \subset \mathcal{R}$ there is a unique least extension a_r, (a_l) which is a right (left) ideal.

Definition 8 A principal right [left] ideal is one of the from $(x)_r[(x_l)]$. The class of all principal right [left] ideals will be denoted by $\bar{R}_{\mathcal{R}}$ $[\bar{L}_{\mathcal{R}}]$.

Lemma 4 *Let $\mathcal{R}$ be a ring, $e \in \mathcal{R}$, then*

- *e is idempotent if and only if $(1-e)$ is idempotent.*
- *$\langle e\rangle_r$ if the set of all x such that $x = ex$ is a principal right ideal.*
- *$\langle e\rangle_r$ and $\langle 1-e\rangle_r$ are mutual inverses, i.e., $\langle e\rangle_r \cup \langle 1-e\rangle_r = 1$, and $\langle e\rangle_r \cap \langle 1-e\rangle_r = 0$.*
- *If $\langle e\rangle_r = \langle f\rangle_r$ and if $\langle 1-e\rangle_r = \langle 1-f\rangle_r$ where e and f are idempotents, then $e = f$.*

Theorem 1 *Two right ideals a and b are inverses if and only if there exists an idempotent e such that $a = \langle e\rangle_r$ and $b = \langle 1-e\rangle_r$.*

Proof Let a and b be inverse right ideals, then there exists elements x, y with $x+y = 1, x \in a, y \in b$. If $z \in a$ then $xz + yz = x$, hence $yz = x - xz$ since $z, xz \in a$ this implies $yz \in a$. But $yz \in b$, hence $yz = 0$. Thus $z = xz \in (x)_r$ for every $z \in a$ and $a \subset (x)_r$. But $x \in a$, hence $a = (x)_r$. Similarly, $b = (y)_r = (1-x)_r$, since $x + y = 1$. Finally, since $z = xz$ for every $z \in a$ this holds for $z = x$ and x is idempotent. □

Theorem 2 *The following statements are equivalent*

(1) Every principal right ideal $\langle a\rangle_r$ has an inverse right ideal.

(2) For every a there exists an idempotent e such that $\langle a\rangle_r = \langle e\rangle_r$.

(3) For every a there exists an element x such that $axa = a$.

(4) For every a there exists an idempotent f such that $\langle a\rangle_l = \langle f\rangle_l$.

(5) Every principal ideal $\langle a\rangle_l$ has an inverse left ideal.

Definition 9 A ring $\mathcal{R}$ is said to be regular if $\mathcal{R}$ possesses anyone of the equivalent properties of the above theorem.

Proposition 1 (cf. [1] Theorem 2.3). *If a and b be two elements in the principal right ideal $\bar{R}_{\mathcal{R}}$ then their join $a \cup b$ is a principal right ideal.*

Also since the principal right ideals $\bar{R}_{\mathcal{R}}$ and the principal left ideals $\bar{L}_{\mathcal{R}}$ are anti-isomorphic it is easy to see that for any two right ideals a and b, we can define $\cap$ as follows:

$$a^l \cap b^l = (a \cup b)^l$$

where a^l is defined by $a^l = \{y : z \in a, \text{implies}, yz = 0\}$ (see [1]).

Lemma 5 (cf. [1] Lemma 2.4). *If a and b be two elements in the principal right ideal $\bar{R}_{\mathcal{R}}$ then $a \cap b$ is a principal right ideal.*

Theorem 3 *The set $\bar{R}_{\mathcal{R}}$ is a complemented modular lattice partially ordered by $\subset$, the meet being $\cap$ and join $\cup$, its zero is (0) and its unit is $(1)_r$.*

1.2 Biordered Sets

Next we recall the definitions of a biordered set from cc. [3]. By a partial algebra E we mean a set together with a partial binary operation on E. The domain of the partial binary operation will be denoted by D_E. On E we define

$$\omega^r = \{(e, f) : fe = e\}\omega^l = \{(e, f) : ef = e\}$$

also, $\mathcal{R} = \omega^r \cap (\omega^r)^{-1}$, $\mathcal{L} = \omega^l \cap (\omega^l)^{-1}$, *and* $\omega = \omega^r \cap \omega^l$

Definition 10 Let E be a partial algebra. Then E is a biordered set if the following axioms and their duals hold:

(1) ω^r and ω^l are quasi-orders on E and

$$D_E = (\omega^r \cup \omega^l) \cup (\omega^r \cup \omega^l)^{-1}.$$

(2) $f \in \omega^r(e) \Rightarrow f\mathcal{R}fe\omega e$.
(3) $g\omega^l f$ and $\quad f, g \in \omega^r(e) \Rightarrow ge\omega^l fe$.
(4) $g\omega^r f\omega^r e \Rightarrow gf = (ge)f$.
(5) $g\omega^l f$ and $f, g \in \omega^r(e) \Rightarrow (fg)e = (fe)(ge)$.

Let $\mathcal{M}(e, f)$ denote the quasi-ordered set $(\omega^l(e) \cap \omega^r(f), <)$ where $<$ is defined by $g < h \Leftrightarrow eg\omega^r eh$, and $gf\omega^l hf$. Then the set

$$S(e, f) = \{h \in M(e, f) : g < h \text{ for all } g \in M(e, f)\}$$

is called the sandwitch set of e and f.

(1) $f, g \in \omega^r(e) \Rightarrow S(f, g)e = S(fe, ge)$

The biordered set E is said to be regular if $S(e, f) \neq \emptyset \; \forall e, f \in E$

Example Let S be a semigroup. On $E(S) = \{e \in S\ e^2 = e\}$ we define

$$e\omega^r f \Leftrightarrow fe = e \text{ and } e\omega^l f \Leftrightarrow ef = e.$$

It can easily be seen that ω^r and ω^l are quasi-orders on $E(S)$. Let $D_{E(S)} = (\omega^r \cup \omega^l) \cup (\omega^r \cup \omega^l)^{-1}$ so whenever $(e, f) \in D_{E(S)}$ we have $ef \in E(S)$. Thus restricting the product in S to $D_E(S)$ we obtain a partial algebra on $E(S)$ and this partial algebra is a biordered set.

2 Biordered Set of a Regular Ring

Let $\mathcal{R}$ be a regular ring and $\bar{R}_{\mathcal{R}}$ be the class of all principal right ideals. For $a \in \bar{R}_{\mathcal{R}}$ there exists an inverse $b \in \bar{R}_{\mathcal{R}}$, by this we mean $a \cap b = 0$ and $a \cup b = 1.$, that is they are complementary. Now for any pair of (a, b) of complementary elements in $\bar{R}_{\mathcal{R}}$, define the mappings $(a, b),\ (a, b)' : \bar{R}_{\mathcal{R}} \to \bar{R}_{\mathcal{R}}$ by

$$(a, b)(x) \mapsto b \cap (a \cup x)$$

and

$$(a, b)'(x) \mapsto a \cup (b \cap x).$$

It is seen that $(a, b)\,[(a, b)']$ is an idempotent order preserving mapping of $\bar{R}_{\mathcal{R}}$ onto the principal ideal $[0, v]\,[n, 1]$.

Let $P(\bar{R}_{\mathcal{R}})$ denote the subsemigroup of $S^*(\bar{R}_{\mathcal{R}})$ generated by these idempotent mappings where the binary composition is defined by

$$(a_1, b_1) \cdot (a_2, b_2) = (a_1 \vee (b_1 \wedge a_2), (b_1 \vee a_2) \wedge b_2)$$

Clearly,

$$E_{P(\bar{R}_{\mathcal{R}})} = \{(a, b) : a, b \in \bar{R}_{\mathcal{R}},\ a \text{ and } b \text{ complementary}\}$$

is a biordered set with quasi-orders defined by

$$(a_1, b_1)\omega^l(a_2, b_2) \text{ if and only if } b_1 \leq b_2$$
$$(a_1, b_1)\omega^r(a_2, b_2) \text{ if and only if } a_2 \leq a_1$$

Lemma 6 *For* $(a_1, b_1)\omega^l(a_2, b_2)$

$$(a_1, b_1)(a_2, b_2) = (a_1, b_1)$$
$$(a_2, b_2)(a_1, b_1) = (a_2 \vee (b_2 \wedge a_1), b_1)$$

Proof We have

$$\begin{aligned}(a_1, b_1) \cdot (a_2, b_2) &= (a_1 \vee (b_1 \wedge a_2), (b_1 \vee a_2) \wedge b_2) \\ &= (a_1, b_1) \\ (a_2, b_2) \cdot (a_1, b_1) &= (a_2 \vee (b_2 \wedge a_1), (b_2 \vee a_1) \wedge b_1) \\ &= (a_2 \vee (b_2 \wedge a_1), b_1)\end{aligned}$$

□

Lemma 7 *For* $(a_1, b_1)\omega^r(a_2, b_2)$

$$\begin{aligned}(a_1, b_1)(a_2, b_2) &= (a_1, b_2 \wedge (a_2 \vee b_1) \\ (a_2, b_2)(a_1, b_1) &= (a_1, b_1)\end{aligned}$$

Theorem 4 $\bar{R}_{\mathcal{R}}$ *be the principal right ideals of the ring* $\mathcal{R}$ *then* $E(\bar{R}_{\mathcal{R}}) = \{(a, b) : a, b \in \bar{R}_{\mathcal{R}}$ *complementary*$\}$, *then* $(E(\bar{R}_{\mathcal{R}}), \omega^r, \omega^l)$ *with* ω^r *and* ω^l *is the biordered set of* $\bar{R}_{\mathcal{R}}$.

References

1. von Neumann, John: Continuous Geometry. Princeton University Press, London (1960)
2. Howie, J.M.: An Introduction to Semigroup Theory. Academic Press, London (1976)
3. Nambooripad, K.S.S.: Structure of regular semigroups-I. Mem. Am. Math. Soc., **22**(224) 1–117 (November 1979) (second of 3 numbers)
4. Pastijn, F.J.: Biordered sets and complemented modular lattices. Semigroup Forum **21**, 205–220 (1980)

Topological Rees Matrix Semigroups

E. Krishnan and V. Sherly

Abstract An important problem in the theory of topological semigroups is to formulate a suitable definition of continuity for the choice of generalized inverses. In this paper, we will show that under certain natural conditions, a topology can be defined on a Rees matrix semigroup, which turns it into a topological semigroup, and in which a canonical continuous choice of inverses is possible. As an example, we show that this construction applied to the semigroup of operators of rank less than or equal to 1 on a Hilbert space gives a topology which is stronger than the norm topology, under which this semigroup is a topological semigroup and the assignment of every operator to its Moore-Penrose inverse is continuous.

Keywords Topological semigroup · Rees matrix semigroup

First we fix the notations and terminology used in this paper. If G is a group and 0 is a symbol not in G, then the semigroup $G^0 = G \cup \{0\}$, with the operation in G extended by $x0 = 0x = 00 = 0$ for all x in G^0, is called a group with zero. Now let X, Y be any two nonempty sets and let P be a function from $X \times Y$ to G^0. Then we can define a multiplication on the set $T = (G \times X \times Y) \cup \{0\}$ by

$$(\alpha, a, b)(\beta, c, d) = \begin{cases} (\alpha P(b, c)\beta, a, d), & \text{if } P(b, c) \neq 0 \\ 0, & \text{if } P(b, c) = 0 \end{cases}$$

$$(\alpha, a, b)0 = 0(\alpha, a, b) = 00 = 0$$

which turns T into a completely 0-simple semigroup, called the Rees $X \times Y$ matrix semigroup over the group with zero G^0 with sandwich matrix P (see [1], Sects. 3.1 and 3.2 and [2], Sect. III.2 for details). This semigroup is denoted $\mathscr{M}^0(G; X, Y; P)$.

E. Krishnan (✉)
University College, Thiruvananthapuram, Kerala, India
e-mail: e.krshnan@gmail.com

V. Sherly
Government College for Women, Thiruvananthapuram, Kerala, India
e-mail: vsherly@gmail.com

P.G. Romeo et al. (eds.), *Semigroups, Algebras and Operator Theory*,
Springer Proceedings in Mathematics & Statistics 142,
DOI 10.1007/978-81-322-2488-4_8

Now suppose that in the above construction, there is a topology on G^0 under which it is a topological semigroup and the mapping $x \mapsto x^{-1}$ is continuous on G. Then G^0 is called a *topological group with zero*. Suppose further that X and Y are also topological spaces and that the mapping $P\colon X \times Y \to G^0$ is continuous, where $X \times Y$ is given the product topology. We show how we can turn T into a topological semigroup in this set-up.

Let $S = G^0 \times X \times Y$ with the product topology and let $Z = \{0\} \times X \times Y$. Then by definition, $T = (S \setminus Z) \cup \{0\}$. Define $\pi\colon S \to G^0$ by $\pi(\lambda, x, y) = \lambda$ and $\pi_0\colon T \to G^0$ by

$$\pi_0(t) = \begin{cases} \pi(t), & \text{if } t \neq 0 \\ 0, & \text{if } t = 0 \end{cases}$$

Note that if $A \subseteq S \setminus Z = S \cap T$, then $\pi(A) = \pi_0(A)$.

We first consider some conditions under which 0 belongs to the closure of $\pi_0(A)$ for a subset A of T. Throughout the sequel, we denote the closure of a subset A of S by $\mathrm{cl}(A)$. The same notation will be used for subsets of G^0 also.

Lemma 1 *Let $A \subseteq T$. If either $0 \in A$ or if $0 \notin A$ and $\mathrm{cl}(A) \cap Z \neq \emptyset$, then $0 \in \mathrm{cl}(\pi_0(A))$. The converse is also true if X and Y are compact.*

Proof If $0 \in A$, then $0 = \pi_0(0) \in \pi_0(A) \subseteq \mathrm{cl}(\pi_0(A))$. Suppose now that $0 \notin A$ and $\mathrm{cl}(A) \cap Z \neq \emptyset$. Since $0 \notin A$, we have $A \subseteq S \setminus Z$ and since $\mathrm{cl}(A) \cap Z \neq \emptyset$, there exists $s \in \mathrm{cl}(A) \cap Z$. Then $s \in \mathrm{cl}(A)$ and $\pi(s) = 0$, by definition of Z and π. Hence $0 = \pi(s) \in \pi(\mathrm{cl}(A))$. Also, since S has the product topology, the projection π is continuous, so that $\pi(\mathrm{cl}(A)) \subseteq \mathrm{cl}(\pi(A))$. Again, since $A \subseteq S \setminus Z$, we have $\pi(A) = \pi_0(A)$. Thus

$$0 \in \pi(\mathrm{cl}(A)) \subseteq \mathrm{cl}(\pi(A)) = \mathrm{cl}(\pi_0(A))$$

Now suppose that X and Y are compact and $0 \in \mathrm{cl}(\pi_0(A))$. If $0 \in A$, we are done. Suppose $0 \notin A$. Since $0 \in \mathrm{cl}(\pi_0(A))$, there exists a net (λ_i) in $\pi_0(A)$ such that $\lim_i \lambda_i = 0$. So for each i, there exists $s_i \in A$ with $\pi_0(s_i) = \lambda_i$. Since $0 \notin A$ we have $A \subseteq S$ and so each $s_i = (\lambda_i, x_i, y_i)$, where $x_i \in X$ and $y_i \in Y$. Now since X and Y are compact, the nets (x_i) and (y_i) have convergent subnets (x_{i_k}) and (y_{i_k}). Let $a = \lim_k x_{i_k}$ and $b = \lim_k y_{i_k}$. Then $((\lambda_{i_k}, x_{i_k}, y_{i_k}))$ is a net in A with $\lim_k(\lambda_{i_k}, x_{i_k}, y_{i_k}) = (0, a, b)$ so that $(0, a, b) \in \mathrm{cl}(A) \cap Z$. Thus $\mathrm{cl}(A) \cap Z \neq \emptyset$. □

The following corollary is immediate

Corollary 2 *If $A \subseteq T$ and $0 \notin \mathrm{cl}(\pi_0(A))$, then $0 \notin A$ and $\mathrm{cl}(A) \cap Z = \emptyset$.* □

We next define an operation on subsets of T which turns out to be a closure operator on T. Let A be a subset of T. If $0 \notin \mathrm{cl}(\pi_0(A))$, then $0 \notin A$ and $\mathrm{cl}(A) \cap Z = \emptyset$, by the corollary above, so that $\mathrm{cl}(A)$ is a subset of T. Also note that for any subset A of T, the set $A \setminus \{0\}$ is contained in S, so that $\mathrm{cl}(A \setminus \{0\})$ is a subset of S and so $(\mathrm{cl}(A \setminus \{0\}) \setminus Z) \cup \{0\}$ is a subset of T.

Thus with each subset A of T, we can associate a subset $\mathsf{c}(A)$ by

$$\mathsf{c}(A) = \begin{cases} \mathrm{cl}(A), & \text{if } 0 \notin \mathrm{cl}(\pi_0(A)) \\ (\mathrm{cl}(A \setminus \{0\}) \setminus Z) \cup \{0\}, & \text{if } 0 \in \mathrm{cl}(\pi_0(A)) \end{cases}$$

We next show that c satisfies the Kuratowski closure axioms (see [3]). In the sequel, for a subset A of T, we often write A' for $A \setminus \{0\}$.

Proposition 3 *The map* c *on subsets of* T *defined above satisfies the following conditions.*

(i) $\mathsf{c}(\emptyset) = \emptyset$
(ii) $A \subseteq \mathsf{c}(A)$, *for all subsets* A *of* T
(iii) $\mathsf{c}(\mathsf{c}(A)) = \mathsf{c}(A)$, *for all subsets* A *of* T.
(iv) $\mathsf{c}(A \cup B) = \mathsf{c}(A) \cup \mathsf{c}(B)$, *for all subsets* A *and* B *of* T.

Proof Since $\pi_0(\emptyset) = \emptyset$, we have $0 \notin \pi_0(\emptyset)$ and so $\mathsf{c}(\emptyset) = \mathrm{cl}(\emptyset) = \emptyset$, by definition of c. This proves (i).

To prove (ii), let $A \subseteq T$ and suppose first that $0 \notin \mathrm{cl}(\pi_0(A))$. Then $\mathsf{c}(A) = \mathrm{cl}(A)$ by definition, and $A \subseteq \mathrm{cl}(A)$, so that $A \subseteq \mathsf{c}(A)$. Next suppose that $0 \in \mathrm{cl}(\pi_0(A))$, so that $\mathsf{c}(A) = (\mathrm{cl}(A') \setminus Z) \cup \{0\}$. Now $A' \subseteq \mathrm{cl}(A')$ and since $A' \subseteq T$, we also have $A' \cap Z = \emptyset$. Hence $A' \subseteq \mathrm{cl}(A') \setminus Z$ and so

$$A = A' \cup \{0\} \subseteq (\mathrm{cl}(A') \setminus Z) \cup \{0\} = \mathsf{c}(A)$$

To prove (iii), let $A \subseteq T$ and let $B = \mathsf{c}(A)$. First suppose that $0 \notin \mathrm{cl}(\pi_0(A))$. Then $B = \mathsf{c}(A) = \mathrm{cl}(A)$, by definition. We next show that $0 \notin \mathrm{cl}(\pi_0(B))$. Since $B = \mathsf{c}(A) \subseteq T$ and $B = \mathrm{cl}(A) \subseteq S$, we have $\pi_0(B) = \pi(B)$. Also, since $B = \mathrm{cl}(A)$ and π is continuous on S, we have

$$\pi(B) = \pi(\mathrm{cl}(A)) \subseteq \mathrm{cl}(\pi(A))$$

Again, since $0 \notin \mathrm{cl}(\pi_0(A))$, we have $0 \notin A$, by Corollary 2 and so $A \subseteq S$. Hence $A \subseteq T \cap S$, so that $\pi(A) = \pi_0(A)$ and so $\mathrm{cl}(\pi(A)) = \mathrm{cl}(\pi_0(A))$.Thus

$$\pi_0(B) = \pi(B) \subseteq \mathrm{cl}(\pi(A)) = \mathrm{cl}(\pi_0(A))$$

and so $\mathrm{cl}(\pi_0(B)) \subseteq \mathrm{cl}(\pi_0(A))$. Since $0 \notin \mathrm{cl}(\pi_0(A))$, by assumption, we also have $0 \notin \mathrm{cl}(\pi_0(B))$. So $\mathsf{c}(B) = \mathrm{cl}(B)$, by definition. Again, since $B = \mathrm{cl}(A)$, we have $\mathrm{cl}(B) = \mathrm{cl}(\mathrm{cl}(A)) = \mathrm{cl}(A)$.Thus $\mathsf{c}(B) = \mathrm{cl}(B) = \mathrm{cl}(A) = B$. In other words, $\mathsf{c}(\mathsf{c}(A)) = \mathsf{c}(A)$. Next suppose that $0 \in \mathrm{cl}(\pi_0(A))$, so that

$$B = \mathsf{c}(A) = (\mathrm{cl}(A') \setminus Z) \cup \{0\}$$

Hence $0 \in B$ so that $0 = \pi_0(0) \in \pi_0(B) \subseteq \mathrm{cl}(\pi_0(B))$ and so by definition, $\mathsf{c}(B) = (\mathrm{cl}(B') \setminus Z) \cup \{0\}$. Now since $B = (\mathrm{cl}(A') \setminus Z) \cup \{0\}$, we have $B' = \mathrm{cl}(A') \setminus Z$

and so $B' \subseteq \text{cl}(A')$. Hence $\text{cl}(B') \subseteq \text{cl}(\text{cl}(A')) = \text{cl}(A')$ and so

$$\mathsf{c}(B) = (\text{cl}(B') \setminus Z) \cup \{0\} \subseteq (\text{cl}(A') \setminus Z) \cup \{0\} = B$$

Since $B \subseteq \mathsf{c}(B)$, by (ii) above, we now have $\mathsf{c}(B) = B$. That is, $\mathsf{c}(\mathsf{c}(A)) = A$.

Finally, let A and B be subsets of T. First note that

$$\text{cl}(\pi_0(A \cup B)) = \text{cl}(\pi_0(A) \cup \pi_0(B)) = \text{cl}(\pi_0(A)) \cup \text{cl}(\pi_0(B)). \quad (1)$$

Now suppose that $0 \notin \text{cl}(\pi_0(A \cup B))$. Then $\mathsf{c}(A \cup B) = \text{cl}(A \cup B)$. Also, by Eq. (1), we have $0 \notin \text{cl}(\pi_0(A))$ and $0 \notin \text{cl}(\pi_0(B))$ so that $\mathsf{c}(A) = \text{cl}(A)$ and $\mathsf{c}(B) = \text{cl}(B)$. Since $\text{cl}(A \cup B) = \text{cl}(A) \cup \text{cl}(B)$, we have

$$\mathsf{c}(A \cup B) = \text{cl}(A \cup B) = \text{cl}(A) \cup \text{cl}(B) = \mathsf{c}(A) \cup \mathsf{c}(B)$$

Next suppose that $0 \in \text{cl}(\pi_0(A \cup B))$. Then

$$\mathsf{c}(A \cup B) = (\text{cl}((A \cup B)') \setminus Z) \cup \{0\}$$

Now

$$(A \cup B)' = (A \cup B) \setminus \{0\} = (A \setminus \{0\}) \cup (B \setminus \{0\}) = A' \cup B'$$

so that

$$\text{cl}((A \cup B)') = \text{cl}(A' \cup B') = \text{cl}(A') \cup \text{cl}(B')$$

and so

$$\text{cl}((A \cup B)') \setminus Z = (\text{cl}(A)' \cup \text{cl}(B')) \setminus Z = (\text{cl}(A') \setminus Z) \cup (\text{cl}(B') \setminus Z)$$

Thus we have

$$\mathsf{c}(A \cup B) = (\text{cl}((A \cup B)') \setminus Z) \cup \{0\} = (\text{cl}(A') \setminus Z) \cup (\text{cl}(B') \setminus Z) \cup \{0\} \quad (2)$$

Again, since $\text{cl}(\pi_0(A \cup B)) = \text{cl}(\pi_0(A)) \cup \text{cl}(\pi_0(B))$ and $0 \in \text{cl}(\pi_0(A \cup B))$, we have $0 \in \text{cl}(\pi_0(A))$ or $0 \in \text{cl}(\pi_0(B))$. Suppose $0 \in \text{cl}(\pi_0(A))$ and $0 \in \text{cl}(\pi_0(B))$. Then $\mathsf{c}(A) = (\text{cl}(A') \setminus Z) \cup \{0\}$ and $\mathsf{c}(B) = (\text{cl}(B') \setminus Z) \cup \{0\}$ so that from Eq. (2)

$$\begin{aligned}\mathsf{c}(A \cup B) &= (\text{cl}(A') \setminus Z) \cup (\text{cl}(B') \setminus Z) \cup \{0\} \\ &= ((\text{cl}(A') \setminus Z) \cup \{0\}) \cup ((\text{cl}(B') \setminus Z) \cup \{0\}) \\ &= \mathsf{c}(A) \cup \mathsf{c}(B)\end{aligned}$$

Next suppose that 0 is in only one of the sets $\text{cl}(\pi_0(A))$ or $\text{cl}(\pi_0(B))$. With a change of notation, if necessary, we can assume that $0 \in \text{cl}(\pi_0(A))$ and $0 \notin \text{cl}(\pi_0(B))$. Then

$$\mathsf{c}(A) = (\mathrm{cl}(A') \setminus Z) \cup \{0\} \quad \text{and} \quad \mathsf{c}(B) = \mathrm{cl}(B)$$

Now since $0 \notin \mathrm{cl}(\pi_0(B))$, we have $0 \notin B$ and $\mathrm{cl}(B) \cap Z = \emptyset$, by Lemma 1. So $B' = B \setminus \{0\} = B$ and hence $\mathrm{cl}(B') \cap Z = \mathrm{cl}(B) \cap Z = \emptyset$. It follows that $\mathrm{cl}(B') \setminus Z = \mathrm{cl}(B') = \mathrm{cl}(B)$. Hence from Eq. (2)

$$\begin{aligned}\mathsf{c}(A \cup B) &= (\mathrm{cl}(A') \setminus Z) \cup (\mathrm{cl}(B') \setminus Z) \cup \{0\}\\ &= ((\mathrm{cl}(A') \setminus Z) \cup \{0\}) \cup (\mathrm{cl}(B))\\ &= \mathsf{c}(A) \cup \mathsf{c}(B)\end{aligned}$$

This proves (iv) and hence the result. □

Since c satisfies the four conditions above, the complements of those subsets F of T with $\mathsf{c}(F) = F$ form a topology on T such that for each subset A of T, the closure of A with respect to this topology is $\mathsf{c}(A)$ (see [3], Chap. 1, Theorem 8). We denote this topology by $\mathscr{T}_{\mathsf{c}}$.

We next look at some of the properties of this topology. First we show that the map π_0 is continuous.

Proposition 4 *The map $\pi_0 \colon T \to G_0$ is continuous with respect to the topology $\mathscr{T}_{\mathsf{c}}$ on T.*

Proof To prove that π_0 is continuous, we need only show that $\pi_0(\mathsf{c}(A)) \subseteq \mathrm{cl}(\pi_0(A))$ for each subset A of T (see [3], Chap. 3, Theorem 1). Let A be a subset of T and first suppose that $0 \notin \mathrm{cl}(\pi_0(A))$. Then $\mathsf{c}(A) = \mathrm{cl}(A)$ and $\mathrm{cl}(A)$ is a subset of S, so that $\pi_0(\mathsf{c}(A)) = \pi_0(\mathrm{cl}(A)) = \pi(\mathrm{cl}(A))$. Now since π is a continuous map on S, we have $\pi(\mathrm{cl}(A)) \subseteq \mathrm{cl}(\pi(A))$. Again, since $A \subseteq S \cap T$, we have $\pi(A) = \pi_0(A)$. Thus in this case.

$$\pi_0(\mathsf{c}(A)) = \pi(\mathrm{cl}(A)) \subseteq \mathrm{cl}(\pi(A)) = \mathrm{cl}(\pi_0(A))$$

Next suppose that $0 \in \mathrm{cl}(\pi_0(A))$, then $\mathsf{c}(A) = (\mathrm{cl}(A') \setminus Z) \cup \{0\}$ so that

$$\pi_0(\mathsf{c}(A)) = \pi_0(\mathrm{cl}(A') \setminus Z) \cup \{\pi_0(0)\} = \pi(\mathrm{cl}(A') \setminus Z) \cup \{0\}$$

since $\mathrm{cl}(A') \setminus Z \subseteq S \setminus Z$ and $\pi_0(0) = 0$. Also, $\pi(\mathrm{cl}(A') \setminus Z) \subseteq \pi(\mathrm{cl}(A'))$ and since π is continuous, we have $\pi(\mathrm{cl}(A')) \subseteq \mathrm{cl}(\pi(A'))$. Again, since $A' \subseteq S \cap T$, we have $\pi(A') = \pi_0(A')$. Hence

$$\pi_0(\mathsf{c}(A)) = \pi(\mathrm{cl}(A') \setminus Z) \cup \{0\} \subseteq \mathrm{cl}(\pi_0(A')) \cup \{0\}$$

Now since $A' \subseteq A$, we have $\pi_0(A') \subseteq \pi_0(A)$ and so $\mathrm{cl}(\pi_0(A')) \subseteq \mathrm{cl}(\pi_0(A))$. Also, $0 \in \mathrm{cl}(\pi_0(A))$, by our assumption. Hence $\mathrm{cl}(\pi_0(A')) \cup \{0\} \subseteq \mathrm{cl}(\pi_0(A))$ and so

$$\pi_0(\mathsf{c}(A)) \subseteq \mathrm{cl}(\pi_0(A')) \cup \{0\} \subseteq \mathrm{cl}(\pi_0(A))$$

Thus $\pi_0(\mathsf{c}(A)) \subseteq \mathrm{cl}(\pi_0(A))$, for each subset A of T and so π_0 is continuous. □

We can also show that the map $\theta\colon S \to T$ defined by

$$\theta(\lambda, x, y) = \begin{cases} (\lambda, x, y), & \text{if } \lambda \neq 0 \\ 0, & \text{if } \lambda = 0 \end{cases}$$

is continuous. For this, we prove that the topology $\mathscr{T}_{\mathsf{c}}$ is weaker than the quotient topology on T induced by θ. First we note a simple property of the map θ.

Lemma 5 *If A is a subset of T, then*

$$\theta^{-1}(A) = \begin{cases} A, & \text{if } 0 \notin A \\ (A \setminus \{0\}) \cup Z, & \text{if } 0 \in A \end{cases}$$

Proof Let A be a subset of T. First, suppose that $0 \notin A$ and let $(\lambda, x, y) \in \theta^{-1}(A)$. Then $\theta(\lambda, x, y) \in A$. Since $0 \notin A$, we have $\theta(\lambda, x, y)) \neq 0$ and so $\theta(\lambda, x, y) = (\lambda, x, y)$, by definition of θ. Thus $(\lambda, x, y) = \theta(\lambda, x, y) \in A$. It follows that $\theta^{-1}(A) \subseteq A$.

To prove the reverse inclusion, let $t \in A$. Then $t \neq 0$, since $0 \notin A$. So, $t = (\lambda, x, y)$ for some $\lambda \in G^0$, $x \in X$ and $y \in Y$. Also, $\lambda \neq 0$, since $t \notin Z$. Hence $\theta(t) = \theta(\lambda, x, y) = (\lambda, x, y) = t$, by definition of θ. Thus $\theta(t) = t \in A$ and so $t \in \theta^{-1}(A)$. It follows that $A \subseteq \theta^{-1}(A)$ also.

Next suppose that $0 \in A$ and let $(\lambda, x, y) \in \theta^{-1}(A)$ so that $\theta(\lambda, x, y) \in A$. If $\lambda = 0$, then $(\lambda, x, y) = (0, x, y) \in Z$ and if $\lambda \neq 0$, then $\theta(\lambda, x, y) = (\lambda, x, y)$, so that $(\lambda, x, y) = \theta(\lambda, x, y) \in A$. Thus $(\lambda, x, y) \in A \cup Z$ and $(\lambda, x, y) \neq 0$, so that $(\lambda, x, y) \in (A \setminus \{0\}) \cup Z$. it follows that $\theta^{-1}(A) \subseteq (A \setminus \{0\}) \cup Z$.

To prove the reverse inclusion, let $A' = A \setminus \{0\}$. Then $A' \subseteq T$ with $0 \notin A'$, so that $\theta^{-1}(A') = A'$, by the first part of the proof. Also, since $A' \subseteq A$, we have $\theta^{-1}(A') \subseteq \theta^{-1}(A)$. Hence $A' = \theta^{-1}(A') \subseteq \theta^{-1}(A)$. Also, $\theta(Z) = \{0\}$, by definition of θ and $0 \in A$, by assumption, so that $\theta(Z) \subseteq A$. Hence $Z \subseteq \theta^{-1}\theta(Z) \subseteq \theta^{-1}(A)$. Since $A' \subseteq \theta^{-1}(A)$ and $Z \subseteq \theta^{-1}(A)$, we have $A' \cup Z \subseteq \theta^{-1}(A)$. □

Now we can compare the topology $\mathscr{T}_{\mathsf{c}}$ and the quotient topology induced by the map θ.

Proposition 6 *The topology $\mathscr{T}_{\mathsf{c}}$ on T is weaker than the quotient topology on T induced by the map $\theta\colon S \to T$, where S has the product topology. If X and Y are compact, then these two topologies are equal.*

Proof Let A be a subset of T which is closed with respect to $\mathscr{T}_{\mathsf{c}}$. We will show that A is closed with respect to $\mathscr{T}_\theta$. Note that since $\mathscr{T}_\theta$ is the quotient topology induced by $\theta\colon S \to T$, the subset A of T is closed with respect to $\mathscr{T}_\theta$ if and only if $\theta^{-1}(A)$ is closed in S.

First suppose that $0 \notin A$. Then $\theta^{-1}(A) = A$, by Lemma 5. Also, $A = \mathsf{c}(A)$, since A is closed with respect to $\mathscr{T}_{\mathsf{c}}$, so that $0 \notin \mathsf{c}(A)$. Also, by definition of c, we have $0 \in \mathsf{c}(A)$, if and only if $0 \in \mathrm{cl}(\pi_0(A))$. It follows that $0 \notin \mathrm{cl}(\pi_0(A))$. Hence $\mathsf{c}(A) = \mathrm{cl}(A)$, by definition of c. Since $\mathsf{c}(A) = A$, we now have $A = \mathrm{cl}(A)$ and so

A is closed in S. Thus $\theta^{-1}(A)$ is closed in S and so A is closed with respect to the quotient topology $\mathscr{T}_\theta$.

Next suppose that $0 \in A$. Then $\theta^{-1}(A) = A' \cup Z$, where $A' = A \setminus \{0\}$, by Lemma 5. Also, in this case, $0 = \pi_0(0) \in \pi_0(A) \subseteq \text{cl}(\pi_0(A))$, so that $\mathsf{c}(A) = (\text{cl}(A') \setminus Z) \cup \{0\}$, by definition of c. Again, since A is closed with respect to $\mathscr{T}_\mathsf{c}$, we have $\mathsf{c}(A) = A$. Thus $A = (\text{cl}(A') \setminus Z) \cup \{0\}$. It follows that $A' = A \setminus \{0\} = \text{cl}(A') \setminus Z$, since $0 \notin \text{cl}(A') \setminus Z$. Hence

$$A' \cup Z = (\text{cl}(A') \setminus Z) \cup Z = \text{cl}(A') \cup Z$$

Now since $\pi\colon S \to G^0$ is continuous and $Z = \pi^{-1}(0)$, we have Z closed in S and so $Z = \text{cl}(Z)$. So,

$$A' \cup Z = \text{cl}(A') \cup \text{cl}(Z) = \text{cl}(A' \cup Z)$$

Thus $A' \cup Z$ is closed in S and since $\theta^{-1}(A) = A' \cup Z$, we have $\theta^{-1}(A)$ closed in S. So, A is closed with respect to $\mathscr{T}_\theta$. Thus every subset of T closed with respect to $\mathscr{T}_\mathsf{c}$ is closed with respect to $\mathscr{T}_\theta$ and so $\mathscr{T}_\mathsf{c}$ is weaker than $\mathscr{T}_\theta$.

Now suppose that X and Y are compact and A is a subset of T which is closed with respect to $\mathscr{T}_\theta$. First suppose that $0 \notin \text{cl}(\pi_0(A))$. Then $\mathsf{c}(A) = \text{cl}(A)$, by definition of c. Also, since A is closed with respect to $\mathscr{T}_\theta$, we have $\theta^{-1}(A)$ closed in S. Again, since $0 \notin \text{cl}(\pi_0(A))$, we have $0 \notin A$, by Corollary 2, so that $\theta^{-1}(A) = A$, by Lemma 5. Hence A is closed in S and so $\text{cl}(A) = A$. Thus $\mathsf{c}(A) = \text{cl}(A) = A$ and so A is closed with respect to $\mathscr{T}_\mathsf{c}$.

Next suppose that $0 \in \text{cl}(\pi_0(A))$, so that $\mathsf{c}(A) = \text{cl}(A') \setminus Z$, where $A' = A \setminus \{0\}$. We now show $0 \in A$. Suppose $0 \notin A$. Then $\theta^{-1}(A) = A$ and $\theta^{-1}(A)$ is closed in S, since A is closed with respect to $\mathscr{T}_\theta$. Thus A is closed in S and so $\text{cl}(A) = A$. On the other hand, since X and Y are compact, the conditions $0 \in \text{cl}(\pi_0(A))$ and $0 \notin A$ imply $\text{cl}(A) \cap Z \neq \emptyset$, by the second part of Lemma 1. This contradicts the earlier conclusion that $\text{cl}(A) = A$, since $A \subseteq T$ and hence $A \cap Z = \emptyset$. Thus $0 \in A$ and so $\theta^{-1}(A) = A' \cup Z$, by Lemma 5. Since $\theta^{-1}(A)$ is closed in S, we have $A' \cup Z$ closed in S and so

$$A' \cup Z = \text{cl}(A' \cup Z) = \text{cl}(A') \cup \text{cl}(Z) = \text{cl}(A') \cup Z$$

since Z is closed in S. Now since $A' \subseteq A \subseteq T$, we have $A' \cap Z = \emptyset$ and so

$$A' = A' \cup Z \setminus Z = \text{cl}(A') \setminus Z$$

Hence

$$A = A' \cup \{0\} = (\text{cl}(A') \setminus Z) \cup \{0\} = \mathsf{c}(A)$$

since $0 \in \text{cl}(\pi_0(A))$, Thus A is closed with respect to $\mathscr{T}_\mathsf{c}$, since $\mathsf{c}(A) = A$. Thus every subset of T closed with respect to $\mathscr{T}_\theta$ is closed with respect to $\mathscr{T}_\mathsf{c}$ and so $\mathscr{T}_\theta$ is weaker than $\mathscr{T}_\mathsf{c}$. Hence in this case, $\mathscr{T}_\mathsf{c} = \mathscr{T}_\theta$. □

Now by definition of quotient topology, the map $\theta\colon S \to T$ is continuous with respect to $\mathscr{T}_\theta$. Since $\mathscr{T}_{\mathsf{c}}$ is weaker than $\mathscr{T}_\theta$, it follows that θ is continuous with respect to $\mathscr{T}_{\mathsf{c}}$ also.

Corollary 7 *The map $\theta\colon S \to T$ is continuous with respect to the product topology on S and the topology $\mathscr{T}_{\mathsf{c}}$ on T.* □

We next show that T with the topology $\mathscr{T}_{\mathsf{c}}$ is a topological semigroup. For this, we first characterize convergence with respect to $\mathscr{T}_{\mathsf{c}}$. Throughout the sequel, T will be assumed to have the topology $\mathscr{T}_{\mathsf{c}}$, and S will be assumed to have the product topology, unless otherwise specified.

Proposition 8 *Let $\{t_i : i \in D\}$ be a net in T. Then we have the following*

(i) $\{t_i : i \in D\}$ converges to 0 in T if and only if the net $\{\pi_0(t_i)\colon i \in D\}$ converges to 0 in G^0.
(ii) For $a \neq 0$ in T, the net $\{t_i : i \in D\}$ converges to a in T if and only if there exists $k \in D$ such that $t_i \neq 0$ for all $i \geq k$ and $\{t_i : i \geq k\}$ converges to a in S.

Proof Let $\{t_i : i \in D\}$ be a net in T. If $\lim_i t_i = 0$, then $\lim_i \pi_0(t_i) = 0$, since π_0 is continuous. Conversely, let $\lim_i \pi_0(t_i) = 0$. Suppose that $\lim_i t_i \neq 0$. Then there exists a neighborhood U of 0 in T such that for each $j \in D$, there exists $k \in D$ with $k \geq j$ such that $t_k \notin U$. Hence the set $E = \{j \in D\colon t_j \notin U\}$ is a cofinal subset of D and so $\{t_j : j \in E\}$ is a subnet of $\{t_i : i \in D\}$. Hence $\{\pi_0(t_j)\colon j \in E\}$ is a subnet of $\{\pi_0(t_i)\colon i \in D\}$. Since the net $\{\pi_0(t_i)\colon i \in D\}$ converges to 0, the subnet $\{\pi_0(t_j)\colon j \in E\}$ also converges to 0. Now if $A = T \setminus U$, then $\{t_j : j \in E\}$ is a net in A, by definition of E and so $\{\pi_0(t_j)\colon j \in E\}$ is a net in $\pi_0(A)$. Since this net converges to 0, we have $0 \in \mathrm{cl}(\pi_0(A))$. Hence $\mathsf{c}(A) = (A \setminus \{0\} \setminus Z) \cup \{0\}$ so that $0 \in \mathsf{c}(A)$. Again, since U is open in T, we have A closed in T so that $\mathsf{c}(A) = A$. Thus $0 \in A$, which contradicts our assumption that $0 \in U = T \setminus A$. It follows that $\lim_i t_i = 0$. This proves (i).

To prove (ii), let $a \in T \setminus \{0\}$ and suppose that $\{t_i : i \in D\}$ converges to a in T. We first show that this net is not frequently 0. Suppose $\{t_i : i \in D\}$ is frequently 0. Then the set $E = \{i \in D\colon t_i = 0\}$ is cofinal in D and so $\{\pi_0(t_j)\colon j \in E\}$ is a subnet of $\{\pi_0(t_i)\colon i \in D\}$. Hence

$$\lim_{j\in E} \pi_0(t_j) = \lim_{i\in D} \pi_0(t_i) = \pi_0(\lim_{i\in D} t_i) = \pi_0(a)$$

since π_0 is continuous. On the other hand, $\lim_{j\in E} t_j = 0$, since $t_j = 0$ for all $j \in E$, so that

$$\lim_{j\in E} \pi_0(t_j) = \pi_0(\lim_{j\in E} t_j) = \pi_0(0) = 0$$

by definition of π_0. It follows that $\pi_0(a) = 0$, since G^0 being a Hausdorff space, nets in it have unique limits. But then $a \in T \setminus \{0\}$, by assumption, so that $\pi_0(a) \neq 0$, by definition of π_0. This contradiction shows that $\{t_i : i \in D\}$ is not frequently 0 and so there exists $k \in D$ with $t_i \neq 0$ for all $i \geq k$.

Next note that $t_i \in T = (S\backslash Z)\cup\{0\}$ and $t_i \neq 0$ for all $i \geq k$, so that $t_i \in S\backslash Z \subseteq S$ for all $i \geq k$. Suppose that $\{t_i : i \geq k\}$ does not converge to a in S. Then there exists a neighborhood U of a in S such that $\{t_i : i \in D\}$ is not eventually in U. Hence the set $E = \{i \in D : i \geq k \text{ and } t_i \notin U\}$ is cofinal in D and so $\{t_j : j \in E\}$ is a subnet of $\{t_i : i \in D\}$. Since the net $\{t_i : i \in D\}$ converges to a in T, so does the subnet $\{t_j : j \in E\}$. Let $F = S \setminus U$ and $A = F \setminus Z$. Then $A \subseteq T$. Also, we have seen that $t_j \in S \setminus Z$ for all $j \in E$ and by definition of E, we have $t_j \notin U$ for all $j \in E$, so that $\{t_j : j \in E\}$ is a net in $(S \setminus Z) \setminus U = (S \setminus U) \setminus Z = A$. Since this net converges to a in T, we have $a \in \mathsf{c}(A)$. Next note that $\mathsf{c}(A) \subseteq \text{cl}(A) \cup \{0\}$, for if $0 \notin \text{cl}(\pi_0(A))$, then $\mathsf{c}(A) = \text{cl}(A)$ and if $0 \in \text{cl}(\pi_0(A))$, then

$$\mathsf{c}(A) = (\text{cl}(A \setminus \{0\}) \setminus Z) \cup \{0\} \subseteq \text{cl}(A \setminus \{0\}) \cup \{0\} \subseteq \text{cl}(A) \cup \{0\}$$

Also, since $A \subseteq F$, we have $\text{cl}(A) \subseteq \text{cl}(F)$. Again, since U is open in S, we have F closed in S so that $\text{cl}(F) = F$. Hence $\text{cl}(A) \subseteq F$ and so $\mathsf{c}(A) \subseteq \text{cl}(A) \cup \{0\} \subseteq F \cup \{0\}$. Since $a \in \mathsf{c}(A)$, we now have $a \in F \cup \{0\}$, which is impossible, since $a \in U = S \setminus F$ and $a \neq 0$. It follows that $\{t_j : j \geq k\}$ converges to a in S.

Conversely, let $\{t_i : i \in D\}$ be a net in T and $a \in T \setminus \{0\}$ such that there exists $k \in D$ with $t_i \neq 0$ for all $i \geq k$ and the net $\{t_i : i \geq k\}$ converges to a in S. then the net $\{\theta(t_i) : i \in D\}$ converges to $\theta(a)$ in T, since $\theta : S \to T$ is continuous. Now since $t_i \in T = (S \setminus Z) \cup \{0\}$ for all $i \in D$ and $t_i \neq 0$ for all $i \geq k$, we have $t_i \in S \setminus Z$ for all $i \geq k$, so that $\theta(t_i) = t_i$ for all $i \geq k$, by definition of θ. Also since $a \in T \setminus \{0\} = S \setminus Z$, we have $\theta(a) = a$. Thus the net $\{t_i : i \geq k\}$ converges to a in T and so does the net $\{t_i : i \in D\}$. □

The above description of convergence in T shows in particular that T is a Hausdorff space.

Corollary 9 *$\mathscr{T}_\mathsf{c}$ is a Hausdorff topology on T.*

Proof It suffices to show that nets in T convergent with respect to $\mathscr{T}_\mathsf{c}$ have unique limits. Let $\{t_i : i \in D\}$ be a net in T which converges to a in T. We will show that this net cannot converge to a different point.

First suppose $a = 0$ and suppose $\{t_i : i \in D\}$ converges to b also in T, where $b \neq a$. First let $a = 0$. Then

$$\lim_{i \in D} \pi_0(t_i) = 0$$

by Proposition 8. Also, $b \neq 0$ and $\{t_i : i \in D\}$ converges to b in T, so that there exists $k \in D$ such that $t_i \neq 0$ for all $i \geq k$ and the net $\{\pi_0(t_i) : i \geq k\}$ converges to b in S, again by Proposition 8. Hence $\lim_{i \geq k} \pi(t_i) = \pi(b)$, since π is continuous on S. Now for each $i \geq k$, we have $t_i \in S \cap T$, so that $\pi(t_i) = \pi_0(t_i)$ and also $b \in S \cap T$, so that $\pi(b) = \pi_0(b)$. Thus

$$\lim_{i \geq k} \pi_0(t_i) = \pi_0(b)$$

Again, since $\{\pi_0(t_i) : i \geq k\}$ is a subnet of $\{\pi_0(t_i) : i \in D\}$, we have

$$\lim_{i \geq k} \pi_0(t_i) = \lim_{i \in D} \pi_0(t_i) = 0$$

Now since G^0 is a Hausdorff space, convergent nets in G^0 have unique limits and so $\pi_0(b) = 0$. But this is impossible, since $b \neq 0$ and so $\pi_0(b) \neq 0$, by definition of π_0. it follows that if $\{t_i : i \in D\}$ converges to 0 in T, then it cannot converge to another nonzero element of T.

Next suppose that $a \neq 0$ and again suppose $\{t_i : i \in D\}$ converges to b also in T, where $b \neq a$. Then there exists $k \in D$ such that $t_i \neq 0$ for all $i \geq k$ and the net $\{t_i : i \geq k\}$ converges to a in S. Also since $\{t_i : i \in D\}$ converges to a and b with $a \neq 0$, we have $b \neq 0$, by the first part of the proof. So there exists $l \in D$ such that $t_i \neq 0$ for all $i \geq l$ and the net $\{t_i : i \geq l\}$ converges to b in S. Now since D is a directed set, there exists $m \in D$ with $m \geq k$ and $m \geq l$. Then the net $\{t_i : i \geq m\}$ is a subnet of both $\{t_i : i \geq k\}$ and $\{t_i : i \geq l\}$ so that $\{t_i : i \geq m\}$ converges to both a and b in S. This is impossible, since S is a Hausdorff space. It follows that $\{t_i : i \in D\}$ cannot have a limit different from a and so is a Hausdorff space. □

Now by a topological semigroup, we mean a semigroup with a Hausdorff topology, with respect to which the multiplication in the semigroup is continuous. We next show that T with the topology $\mathscr{T}_c$ is a topological semigroup. For this, we first note that S is a semigroup with respect to multiplication defined by, under a simple condition on the function P

$$(\alpha, a, b) \cdot (\beta, c, d) = (\alpha P(b, c)\beta, a, d)$$

(see [1], Sect. 3.1). We can show that S is a topological semigroup with respect to the product topology. First note that the product topology on $S = G^0 \times X \times Y$ is Hausdorff, since G^0, X and Y are Hausdorff spaces. Also, if $\{(\lambda_i, x_i, _i, y_i) : i \in D\}$ is a net in S which converges to (α, a, b), then $\lim_i \lambda_i = \alpha$, $\lim_i x_i = a$ and $\lim_i y_i = b$, by the characterization of convergence in a product space ([3], Chap. 3, Theorem 4).

Suppose $\{(\lambda_i, x_i, y_i) : i \in D\}$ and $\{(\mu_i, u_i, v_i) : i \in D\}$ are nets in S converging to (α, a, b) and (β, c, d_i). Then for each i, we have $(\lambda_i, x_i, y_i) \cdot (\mu_i, u_i, v_i) = (\lambda_i P(y_i, u_i)\mu_i, x_i, v_i)$. By the description of convergence in S, we have $\lim_i \lambda_i = \alpha$ and $\lim_i \mu_i = \beta$. Also $\lim_i y_i = b$ and $\lim_i u_i = c$ so that $\lim_i P(y_i, u_i) = P(b, c)$, since P is continuous. Hence $\lim_i \lambda_i P(y_i, u_i)\mu_i = \alpha P(b, c)\beta$, since multiplication in G^0 is continuous. Again, from the convergence of the nets in S, we have $\lim_i x_i = a$ and $\lim_i v_i = d$. So, the net $\{(\lambda_i P(y_i, u_i)\mu_i, x_i, v_i) : i \in D\}$ converges to $(\alpha P(b, c)\beta, a, d)$. It follows that S is a topological semigroup.

We next note that the map θ is a homomorphism from S to T. Let (α, a, b) and (β, c, d) be elements of S. First assume that both α and β are nonzero. Then by definition of θ, we have $\theta(\alpha, a, b) = (\alpha, a, b)$ and $\theta(\beta, c, d) = (\beta, c, d)$, so that

$$\theta(\alpha, a, b)\theta(\beta, c, d) = (\alpha, a, b)(\beta, c, d) = \begin{cases} (\alpha P(b, c)\beta, a, d), & \text{if } P(b, c) = 0 \\ 0, & \text{if } P(b, c) \neq 0 \end{cases}$$

by definition of the multiplication in T and

$$\theta((\alpha, a, b)\cdot(\beta, c, d)) = \theta(\alpha P(b, c)\beta, a, d) = \begin{cases} (\alpha P(b, c)\beta, a, d), & \text{if } \alpha P(b, c)\beta = 0 \\ 0, & \text{if } \alpha P(b, c) \neq \beta 0 \end{cases}$$

by definition of the product in S and by definition of θ. Now since α and β are both nonzero elements of G^0, the product $\alpha P(b, c)\beta = 0$ iff $P(b, c) = 0$. It follows that $\theta((\alpha, a, b)\cdot(\beta, c, d)) = \theta(\alpha, a, b)\theta(\beta, c, d)$ in this case. Next suppose that α or β is equal to 0. Then $\alpha P(b, c)\beta = 0$ so that

$$\theta((\alpha, a, b)\cdot(\beta, c, d)) = \theta(\alpha P(b, c), a, d) = \theta(0, a, d) = 0$$

Again, $\theta(\alpha, a, b)$ or $\theta(\beta, c, d)$ is 0 (depending on whether α or β is 0) and so $\theta(\alpha, a, b)\theta(\beta, c, d) = 0$. Thus $\theta((\alpha, a, b)\cdot(\beta, c, d)) = \theta(\alpha, a, b)\theta(\beta, c, d)$ in this case also. Thus θ is a homomorphism.

We need to impose a condition on the map P for T to be a topological semigroup with respect to $\mathscr{T}_\mathsf{C}$.

Definition 10 Let G^0 be a topological group with zero and let X be a set. A map $\phi\colon X \to G^0$ is said to be *bounded*, if for any net $\{g_i : i \in D\}$ in G^0 converging to 0 and for any net $\{x_i : i \in D\}$ in X, the net $\{g_i\phi(x_i)\colon i \in D\}$ converges to 0 in G^0.

Note that if $\mathbb{K}$ is the set of real numbers or the set of complex numbers with multiplication, then $\phi\colon X \to \mathbb{K}$ is bounded in the usual sense iff ϕ is bounded in the above sense.

Theorem 11 *If the map $P\colon X \times Y \to G^0$ is bounded, then T is a topological semigroup with respect to the topology $\mathscr{T}_\mathsf{C}$.*

Proof We have seen in Corollary 9 that $\mathscr{T}_\mathsf{C}$ is a Hausdorff topology on T. To prove that the multiplication in T is continuous, let $\{s_i : i \in D\}$ and $\{t_i : i \in D\}$ be nets converging to a and b in T. We will show that $\{s_it_i : i \in D\}$ converges to ab in T.

First suppose that both a and b are nonzero. Then there exists $k \in D$ such that $s_i \neq 0$ for all $i \geq k$ and $\{s_i : i \geq k\}$ converges to a in S and also there exists $l \in D$ such that $t_i \neq 0$ for all $i \geq l$ and $\{t_i : i \geq l\}$ converges to b in S. Since D is a directed set, there exists $m \in D$ such that $m \geq k$ and $m \geq l$. Then $s_i \neq 0$ for all $i \geq m$ and $\{s_i : i \geq m\}$ is a subnet of $\{s_i : i \geq k\}$, since $m \geq k$. Hence $\{s_i : i \geq m\}$ also converges to a. Again, $t_i \neq 0$ for all $i \geq m$ and $\{t_i : i \geq m\}$ converges to b, since $m \geq l$. Now since S is a topological semigroup, the product net $\{s_i \cdot t_i : i \geq m\}$ converges to $a \cdot b$ in S. Hence the net $\{\theta(s_i \cdot t_i)\colon i \geq m\}$ converges to $\theta(a \cdot b)$ in T, since $\theta\colon S \to T$ is continuous. Also, since θ is a homomorphism, we have

$$\theta(s_i \cdot t_i) = \theta(s_i)\theta(t_i) \quad \text{for all } i \in D$$

and

$$\theta(a \cdot b) = \theta(a)\theta(b)$$

Again, since s_i and t_i are nonzero for all $i \geq k$, we have

$$\theta(s_i) = s_i \quad \text{and} \quad \theta(t_i) = t_i \quad \text{for all } i \geq k,$$

by definition of θ. Also, since a and b are nonzero, we have $\theta(a) = a$ and $\theta(b) = b$. Thus the net $\{s_it_i : i \geq m\}$ converges to ab in T, and hence $\{s_it_i : i \in D\}$ also converges to ab in T.

Next suppose that a or b is 0. By a change of notation, if necessary, we can assume that $a = 0$. If now $\{s_it_i : i \in D\}$ is eventually equal to 0, then this net evidently converges to 0 in T. Suppose this net is not eventually equal to 0. Then the set $E = \{i \in D\colon s_it_i \neq 0\}$ is cofinal in D and so $\{s_jt_j\colon j \in E\}$ is subnet of $\{s_it_i : i \in D\}$. Now for each $j \in E$, we have $s_jt_j \neq 0$ and so $s_j \neq 0$ and $t_j \neq 0$. Let $s_j = (\lambda_j, x_j, y_j)$ and $t_j = (\mu_j, u_j, v_j)$ for each $j \in E$. Then $s_jt_j = (\lambda_j P(y_j, u_j)\mu_j, x_j, v_j)$, by definition of the product in T and so

$$\pi_0(s_jt_j) = \lambda_j P(y_j, u_j)\mu_j$$

by definition of π_0. Now $\{s_i : i \in D\}$ converges to $a = 0$ and the map π_0 is continuous by Proposition 4 so that we have $\lim_{i\in D} \pi_0(s_i) = 0$. Also, $\{\pi_0(s_j)\colon j \in E\}$ is a subnet of $\{\pi_0(s_i)\colon i \in D\}$, since E is cofinal in D. Hence $\lim_{j\in E} \lambda_j = \lim_{j\in E} \pi_0(s_j) = 0$ and so

$$\lim_{j\in E} \lambda_j P(y_j, u_j) = 0$$

since P is a bounded map to G^0. Also, since $\{s_j\colon j \in E\}$ is a subnet of $\{s_i : i \in D\}$ and the latter net converges to b, so does the former net and so

$$\lim_{j\in E} \mu_j = \lim_{j\in} \pi_0(t_j) = \pi_0(b)$$

since π_0 is a continuous map on T. So,

$$\lim_{j\in E} \pi_0(s_jt_j) = \lim_{j\in E} (\lambda_j P(y_j, u_j))\mu_j = 0\,\pi_0(b) = 0$$

since multiplication is continuous in G^0. it follows that $\{s_jt_j\colon j \in E\}$ converges to 0 in T. To show that $\{s_it_i : i \in D\}$ also converges to 0 in T, let U be a neighborhood of 0 in T. Since $\{s_jt_j\colon j \in E\}$ converges to 0, there exists $k \in E$ such that $s_jt_j \in U$ for all $j \in E$ with $j \geq k$. Let $i \in D$ with $i \geq k$. If $i \in E$, then $s_it_i \in U$. If $i \notin E$, then $s_it_i = 0$, by definition of E and so $s_it_i = 0 \in U$. Thus $s_it_i \in U$ for all $i \in D$ with $i \geq k$. It follows that $\{s_it_i : i \in D\}$ converges to 0 in T.

Thus in all cases, $\{s_it_i : i \in D\}$ converges to ab. It follows that multiplication in T is continuous, and hence T is a topological semigroup. □

It is well known that a Rees matrix semigroup $\mathscr{M}^0(G; X, Y; P)$ is regular if and only if for each $x \in X$ there exists a $y \in Y$ with $P(x, y) \neq 0$ and for each $y \in Y$ there exists $x \in X$ with $P(x, y) \neq 0$ ([1], Lemma 3.1). For this reason, P is called *regular* if it satisfies the above condition. So, in the above result, if P is also assumed to be regular, then T is a topological semigroup which is also regular.

Now one problem in topological semigroups which are regular is the correct formulation of the connection between the topology and the notion of regularity. Since in a regular semigroup, inversion is a relation (rather than a function as in the case of a group), it is not very clear how the continuity of inversion is to be formulated so as to be meaningful in the significant examples. In many cases, there is no canonical choice of a single inverse.

However, in certain types of Rees matrix semigroups, such a canonical choice of inverses is possible. Consider a Rees matrix semigroup $T = \mathscr{M}^0(G; X, X; P)$ and suppose that for each x in X, we have $P(x, x) = \epsilon$, where ϵ is the identity of G. Then for each (α, a, b) in T, we can define $(\alpha, a, b)^\dagger = (\alpha^{-1}, b, a)$. Then $(\alpha, a, b)^\dagger$ is an inverse of (α, a, b), for

$$\begin{aligned}(\alpha, a, b)(\alpha, a, b)^\dagger(\alpha, a, b) &= (\alpha, a, b)(\alpha^{-1}, b, a)(\alpha, a, b)\\ &= (\alpha P(b, b)\alpha^{-1}, a, a)(\alpha, a, b)\\ &= (\epsilon, a, a)(\alpha, a, b)\\ &= (\epsilon P(a, a), a, b)\\ &= (\alpha, a, b)\end{aligned}$$

using the fact that $P(b, b) = P(a, a) = \epsilon$. Also,

$$\begin{aligned}(\alpha, a, b)^\dagger(\alpha, a, b)(\alpha, a, b)^\dagger &= (\alpha^{-1}, b, a)(\alpha, a, b)(\alpha^{-1}, b, a)\\ &= (\alpha^{-1}P(a, a)\alpha, b, b)(\alpha^{-1}, b, a)\\ &= (\epsilon, b, b)(\alpha^{-1}, b, a)\\ &= (\epsilon P(b, b)\alpha^{-1}, b, a)\\ &= (\alpha^{-1}, b, a)\\ &= (\alpha, a, b)^\dagger\end{aligned}$$

We can also show that if in addition, X is a topological space, G^0 is a topological group with zero and P is continuous and bounded, then the above choice of inverses for nonzero elements is continuous. To see this, let $\{t_i : i \in D\}$ be a net in $T \setminus \{0\}$ which converges to p in $T \setminus \{0\}$. Then $t_i = (\lambda_i, x_i, y_i)$ with $\lambda_i \in G$, $x_i \in X$ and $y_i \in Y$ for each $i \in D$ and $p = (\alpha, a, b)$ with $\alpha \in G$, $a \in X$ and $b \in Y$. Since $t_i \in T \setminus \{0\} = S \setminus Z$ and $t \neq 0$, the net $\{t_i : i \in D\}$ converges to p in S, by Proposition 8. Hence $\lim_i \lambda_i = \lambda$, $\lim_i x_i = a$, and $\lim_i y_i = b$, since the topology on S is the product topology. Since $\lim_i \lambda_i = \alpha$ in G and G^0 is a topological group with zero, we have $\lim_i \lambda_i^{-1} = \alpha^{-1}$. Hence $\{(\lambda_i^{-1}, x_i, y_i) : i \in D\}$ converges to (α^{-1}, a, b) in S. In other words $\{{t_i}^\dagger : i \in D\}$ is a net in T such that ${t_i}^\dagger \neq 0$ for each $i \in D$ and converging to $p^\dagger$ in S. So, $t_i^\dagger : i \in D\big\}$ converges to $p^\dagger$ in T. It follows that the map $t \mapsto t^\dagger$ is continuous in $T \setminus \{0\}$. Thus we have the following result:

Theorem 12 *Let G^0 be a group with zero 0, X be a set and P be a map from $X \times X$ to G^0. If $P(x, x) = \epsilon$, where ϵ is the identity of G, for each $x \in X$, then the Rees matrix* *semigroup* $T = \mathscr{M}(G; X, X; P)$ *is a regular semigroup and for each* (α, a, b) *in* T, *the element* $(\alpha, a, b)^\dagger$ *defined by* $(\alpha, a, b)^\dagger = (\alpha^{-1}, b, a)$ *is an inverse of* (α, a, b).

Moreover if G^0 *is a topological group with zero,* X *is a topological space and* P *is continuous and bounded, then* T *is a topological semigroup and the map* $t \mapsto t^\dagger$ *is a continuous choice of inverses for nonzero elements of* T. □

As an example of this, we consider the semigroup $\mathcal{K}_1(\mathsf{H})$ of all operators of rank 1 or less on a Hilbert space H. It is not difficult to show that for any two vectors a and b of H, the operator $a \otimes b$ on H defined by

$$(a \otimes b)(x) = \langle x, b\rangle a$$

is in $\mathcal{K}_1(\mathsf{H})$ and that for every operator t in $\mathcal{K}_1(\mathsf{H})$, there exists two *unit* vectors a and b in H and a nonzero complex number α such that $t = \alpha(a \otimes b)$. Using this, we can prove the following result (see [4], Theorem 2.8).

Proposition 13 *Let* V *be the subset of the Hilbert space* H *consisting of exactly one unit vector from each one-dimensional subspace of* H. *Let* $\mathbb{C}^*$ *be the multiplicative group of nonzero complex numbers and let* P *be defined on* $V \times V$ *by* $P(x, y) = \langle y, x\rangle$, *the inner product in* H. *Then the semigroup* $\mathcal{K}_1(H)$ *of operators of rank one or less on* H *is isomorphic with the Rees matrix semigroup* $T = \mathscr{M}^0(\mathbb{C}^*; V, V; P)$ *under the map* ϕ *on* T *defined by* $\phi(\alpha, a, b) = \alpha(a \otimes b)$ *and* $\phi(0) = 0$. □

Also, every finite rank operator t on H h as a unique generalized inverse $t^\dagger$ (called the Moore-Penrose inverse of t) such that $tt t^\dagger$ and $t^\dagger t$ are self-adjoint idempotents ([4], Theorem 1.1, Theorem 1.4). It is not difficult to prove that that in the above construction, $\phi((\alpha, a, b)^\dagger) = \phi(\alpha, a, b)^\dagger$

In the semigroup $T = \mathscr{M}^0(\mathbb{C}^*; V, V; P)$ constructed above, $\mathbb{C}^*$ is a topological group with zero in the sense of our definition, since multiplication is continuous in $\mathbb{C}$ and inversion is continuous in $\mathbb{C}^*$. Also, the set V consisting of exactly one unit vector from each one-dimensional subspace of the Hilbert space H has the relative topology induced by the norm topology on H. The map P is continuous, since the inner product is continuous on $H \times H$. It is also bounded, for if x and y are in V, then $||x|| = ||y|| = 1$, so that by the Cauchy-Schwartz inequality, $|\langle y, x\rangle| \leq ||x||\,||y|| = 1$. It now follows that T is a topological semigroup with respect to the topology $\mathscr{T}_c$ described above.

Also we have noted that the map ϕ in the last theorem above is an isomorphism onto the semigroup $\mathcal{K}_1(H)$. Hence with respect to the topology $\mathscr{T}_c$ on T and the quotient topology $\mathscr{T}_\phi$ induced by ϕ on T, the map ϕ is a homeomorphism of T onto $\mathcal{K}_1(H)$. Hence $\mathcal{K}_1(H)$ is a topological semigroup with respect to the topology $\mathscr{T}_\phi$. We can show that the $t \mapsto t^\dagger$ is continuous on nonzero operators in $\mathcal{K}_1(H)$ with respect to this topology:

Proposition 14 *The map* $t \mapsto t^\dagger$, *where* $t^\dagger$ *is the Moore-Penrose inverse of* t, *is continuous on* $\mathcal{K}_1(H) \setminus \{0\}$ *with respect to the topology* $\mathscr{T}_\phi$

Proof Let t be a nonzero element of $\mathcal{K}_1(H)$ and let (s_i) be a net in $\mathcal{K}_1(H)$ converging to t with respect to the topology $\mathscr{T}_\phi$. Then the net $(\phi^{-1}(s_i))$ converges to $\phi^{-1}(t)$ with respect to the topology $\mathscr{T}_\mathsf{C}$; and so by Proposition 8, there exists k such that $\phi^{-1}(s_i) \neq 0$ for all $i \geq k$ and $\phi^{-1}(s_i)$ converges to $\phi^{-1}(t)$ in $S = \mathbb{C} \times V \times V$ with respect to the product topology. Now since $\phi^{-1}(s_i) \neq 0$ for $i \geq k$, there exist λ_i in $\mathbb{C}$ and (x_i), (y_i) in V such that $\phi^{-1}(s_i) = (\lambda_i, x_i, y_i)$ for each $i \geq k$. Again, since $\phi^{-1}(t) \neq 0$, there exist α in $\mathbb{C}$ and a, b in V such that $\phi^{-1}(t) = (\alpha, a, b)$. Then $((\lambda_i, x_i, y_i))$ converges to (α, a, b) in S with respect to the product topology, so that (λ_i) converges to α in $\mathbb{C}$ and (x_i), (y_i) converges to a, b, respectively, in V. Now since $s_i = \phi(\lambda_i, x_i, y_i) = \lambda_i(x_i \otimes y_i)$, we have $s_i{}^\dagger = \lambda^{-1}(y_i \otimes x_i)$ and since $t = \phi(\alpha, a, b) = \alpha(a \otimes b)$, we have $t^\dagger = \alpha^{-1}(b \otimes a)$. Hence $\phi^{-1}(s_i{}^\dagger) = (\lambda_i^{-1}, y_i, x_i)$ and $\phi^{-1}(t^\dagger) = (\alpha^{-1}, b, a)$. So,

$$
\begin{aligned}
\lim_{i \geq k} \phi^{-1}(s_i{}^\dagger) &= \lim_{i \geq k}(\lambda_i^{-1}, y_i, x_i) \\
&= (\lim_{i \geq k} \lambda_i^{-1}, \lim_{i \geq k} y_i, \lim_{i \geq k} x_i) \\
&= (\alpha^{-1}, b, a) \\
&= \phi^{-1}(t^\dagger)
\end{aligned}
$$

using the fact that inversion is continuous on nonzero complex numbers. It follows that $(\phi^{-1}(s_i{}^\dagger))$ converges to $\phi^{-1}(t^\dagger)$ in T with respect to $\mathscr{T}_\mathsf{C}$ and hence $(s_i{}^\dagger)$ converges to $t^\dagger$ in $\mathcal{K}_1(H)$ with respect to $\mathscr{T}_\phi$.

Thus for every net (s_i) converging to t with respect to $\mathscr{T}_\phi$ in $\mathcal{K}_1(H)$, the net $(s_i{}^\dagger)$ converges to $t^\dagger$ with respect to $\mathscr{T}_\phi$ and so the map $t \mapsto t^\dagger$ is continuous with respect to $\mathscr{T}_\phi$. □

References

1. Clifford, A.H., Preston, G.B.: The algebraic theory of semigroups, Math. Surveys No.7., Amer. Math. Soc., Providence, (Vol. I) 1961, (Vol. II) (1967)
2. Howie, J.M.: An Introduction to Semigroup Theory. Academic Press, London (1976)
3. Kelley, J.L.: General Topology. Springer, New York (1955)
4. Krishnan, E., Sherly, V.: Semigroup of finite rank operators. Bull. Cal. Math. Soc. **82**, 223–240 (1990)

Prime Fuzzy Ideals, Completely Prime Fuzzy Ideals of Po-Γ-Semigroups Based on Fuzzy Points

Pavel Pal, Sujit Kumar Sardar and Rajlaxmi Mukherjee Pal

Abstract Using fuzzy points the notions of prime fuzzy ideals, weakly prime fuzzy ideals, completely prime fuzzy ideals, and weakly completely prime fuzzy ideals of a po-Γ-semigroup have been introduced. Some important properties and characterizations of these ideals have been obtained. The relations among various types of primeness have also been investigated.

Keywords Γ-semigroup, po-Γ-semigroup · Fuzzy point · Fuzzy ideal · Prime fuzzy ideal · Weakly prime fuzzy ideal · Completely prime fuzzy ideal · Weakly completely prime fuzzy ideal

AMS Subject Classification (2010) 08A72 · 20M12 · 20M99

1 Introduction

Uncertainty is an attribute of information and uncertain data are presented in various domains. The most appropriate theory for dealing with uncertainties was introduced by Zadeh [28] in 1965 by defining fuzzy set which has opened up keen insights and applications in vast range of scientific fields. Rosenfeld [16] pioneered the study of fuzzy algebraic structures by introducing the notions of fuzzy groups and showed that many results in groups can be extended in an elementary manner to develop algebraic concepts. After that Kuroki [12, 14] started the study of fuzzy ideal theory in

---

Author is grateful to University Grants Commission, Govt. of India, for providing research support as SRF.

P. Pal (✉) · S.K. Sardar · R. Mukherjee Pal
Department of Mathematics, Jadavpur University, Kolkata 700032, India
e-mail: ju.pavel86@gmail.com

S.K. Sardar
e-mail: sksardarjumath@gmail.com

R. Mukherjee
e-mail: ju.rajlaxmi@gmail.com

P.G. Romeo et al. (eds.), *Semigroups, Algebras and Operator Theory*,
Springer Proceedings in Mathematics & Statistics 142,
DOI 10.1007/978-81-322-2488-4_9

semigroups. Xie [26] used the notion of fuzzy points to introduce prime fuzzy ideals in semigroups. The notion of Γ-semigroups was introduced by M.K. Sen [22] as a generalization of semigroups. T.K. Dutta and N.C. Adhikari [5] developed the theory of Γ-semigroups by introducing the notion of operator semigroups. Γ-semigroups have also been the object of study of many researchers like Chattopadhyay [1, 8], Chinram et al. [2]. The notion of Γ-semigroups has been extended to fuzzy setting by S.K. Sardar and S.K. Majumder [17–19]. They have studied fuzzy ideals, fuzzy prime ideals, fuzzy semiprime ideals, and fuzzy ideal extensions in Γ-semigroups directly as well as via operator semigroups. Sen and Seth [24] introduced the notion of po-Γ-semigroups. Among the other papers of po-Γ-semigroups we refer to [7, 25]. Kehayopulu has contributed a lot to the ordered semigroups by using fuzzy notion [9, 10]. In this paper we investigate in po-Γ-semigroups the validity of various properties of prime fuzzy ideals of semigroups [26], Γ-semigroups [18, 21] as well as of po-semigroups [10, 27]. We study here prime fuzzy ideals, weakly prime fuzzy ideals, completely prime fuzzy ideals, and weakly completely prime fuzzy ideals in po-Γ-semigroups by using the notion of fuzzy points.

It is important to mention here as to why different types of prime ideals arise in fuzzy setting in contrast with the crisp setting of semigroups or Γ-semigroups. When we formulate some fuzzy notions, to check the correctness of the formulation, we always verify whether the level subset criterion and characteristic function criterion are satisfied. Some situations are very nice where translations of crisp notions to fuzzy setting become compatible with the level subset criterion and characteristic function criterion. But in case of prime fuzzy ideals the situation is not so nice. Just by analogy with the definition of prime ideal in crisp algebra (*cf.* Definition 4.1) if we define prime fuzzy ideal (*cf.* Definition 4.3) in po-Γ-semigroups then we see that level subset criterion does not hold (*cf.* Example 4.13). In order to make the notion compatible with the level subset criterion (*cf.* Theorem 4.18) the notion of weakly prime fuzzy ideal (*cf.* Definition 4.17) is introduced.

We organize the paper as follows. In Sect. 2 we recall some preliminary notions of po-Γ-semigroups as well as of fuzzy subsets. In Sect. 3 we define fuzzy points and their composition in a po-Γ-semigroup and subsequently characterize composition of two fuzzy points in po-Γ-semigroups (*cf.* Theorem 3.2). Also some related properties of fuzzy points are studied in this section. In Sect. 4 prime fuzzy ideals of po-Γ-semigroups are defined. We then obtain various properties of prime fuzzy ideals (*cf.* Proposition 4.8, Theorems 4.7, 4.10, Corollary 4.11). An important characterization of prime fuzzy ideals is also obtained (*cf.* Theorem 4.15). Weakly prime fuzzy ideals of Γ-semigroups are then defined and studied. It is shown that unlike prime fuzzy ideals they satisfy level subset criterion (*cf.* Theorem 4.18). Some other important properties of weakly prime fuzzy ideals are also obtained (*cf.* Theorem 4.23). In Sect. 5 we introduce the notion of completely prime fuzzy ideals (*cf.* Definition 5.1) and weakly completely prime fuzzy ideals (*cf.* Definition 5.2) in po-Γ-semigroups and study their properties (*cf.* Theorems 5.3, 5.5–5.7).

2 Preliminaries

In this section we recall some elementary notions for their use in the sequel.

Definition 2.1 ([23]) Let $S = \{x, y, z, \ldots\}$ and $\Gamma = \{\alpha, \beta, \gamma \ldots\}$ be two nonempty sets. Then S is called a Γ-semigroup if there exists a mapping $S \times \Gamma \times S \rightarrow S$, written as $(a, \alpha, b) \rightarrow a\alpha b$ satisfying (1) $x\gamma y \in S$, (2) $(x\beta y)\gamma z = x\beta(y\gamma z)$, for all $x, y, z \in S$, $\alpha, \beta, \gamma \in \Gamma$.

Remark 2.2 Definition 2.1 is the definition of one-sided Γ-semigroup. It may be noted here that in 1981, Sen [22] introduced the notion of both-sided Γ-semigroups.

Example 2.3 ([22]) Let S be the set of all 2×3 matrices over the set of positive integers and Γ be the set of all 3×2 matrices over same set. Then S is a both-sided as well as a one-sided Γ-semigroup with respect to the usual matrix multiplication.

The following example shows that there exists a one-sided Γ-semigroup which is not a both-sided Γ-semigroup.

Example 2.4 ([1]) Let S be a set of all negative rational numbers. Obviously, S is not a semigroup under usual product of rational numbers. Let $\Gamma = \{-\frac{1}{p}\colon \text{p is prime}\}$. Let $a, b, c \in S$ and $\alpha, \beta \in \Gamma$. Now if $a\alpha b$ is equal to the usual product of rational numbers a, α, b then $a\alpha b \in S$ and $(a\alpha b)\beta c = a\alpha(b\beta c)$. Hence S is a one-sided Γ-semigroup. It is also clear that it is not a both-sided Γ-semigroup.

Definition 2.5 ([24]) A Γ-semigroup S is said to be a po-Γ-semigroup (partially ordered Γ-semigroup) if (1) $(S, \leq)$ and $(\Gamma, \leq)$ are posets, (2) $a \leq b$ in S implies that $a\alpha c \leq b\alpha c$, $c\alpha a \leq c\alpha b$ in S and $\alpha \leq \beta$ in Γ implies $a\alpha b \leq a\beta b$ in S for all $a, b, c \in S$ and for all $\alpha, \beta \in \Gamma$.

Example 2.6 ([24]) Let S be the set of all isotone mappings from a poset P to another poset Q and Γ be the set of all isotone mappings from a poset Q to another poset P. Let $f, g \in S$ and $\alpha \in \Gamma$. Then $f\alpha g$ denotes the usual mapping composition of f, α and g. The relation $\leq$ on S defined by $f \leq g$ if and only if $f(a) \leq g(a)$ for all $a \in P$ is a partial order on S. In a similar fashion Γ can be made into a poset. It can be verified that S is a po-Γ-semigroup.

Remark 2.7 Definition 2.5 is the definition of one-sided po-Γ-semigroups. It may be noted that the definition of both-sided po-Γ-semigroups [7] was introduced by T.K. Dutta and N.C. Adhikari. Throughout this paper unless otherwise mentioned S, a po-Γ-semigroup, is considered to be one sided.

Definition 2.8 A po-Γ-semigroup S is called a commutative po-Γ-semigroup if $a\alpha b = b\alpha a$, for all $a, b \in S$ and $\alpha \in \Gamma$.

Definition 2.9 ([13]) Let S be a po-Γ-semigroup. A nonempty subset I of S is said to be a right ideal (left ideal) of S if (1) $I\Gamma S \subseteq I$ (resp. $S\Gamma I \subseteq I$), (2) $a \in I$ and $b \leq a$ imply $b \in I$. I is said to be an ideal of S if it is a right ideal as well as a left ideal of S.

Definition 2.10 ([13]) Let A be a subset of a po-Γ semigroup S. Then we define $(A] := \{x \in S : x \leq y \text{ for some } y \in A\}$. If A is a singleton $\{a\}$, then for simplification we write $(a]$ instead of $(\{a\}]$.

Proposition 2.11 ([11]) *Let S be a po-Γ-semigroup, A and B be two nonempty subsets of S. Then $(A]\Gamma(B] \subseteq (A\Gamma B]$.*

Moreover, if A and B are any two ideals (left, right or both sided) of S, then

(1) $(A] = A, (B] = B$,
(2) $(A \cap B] = (A] \cap (B]$, *and*
(3) $(A \cup B] = (A] \cup (B]$.

Definition 2.12 ([28]) A fuzzy subset μ of a nonempty set X is a function $\mu : X \to [0, 1]$.

Definition 2.13 ([4]) Let μ be a fuzzy subset of a nonempty set X. Then the set $\mu_t = \{x \in X : \mu(x) \geq t\}$ for $t \in [0, 1]$, is called the level subset or t-level subset of μ.

Definition 2.14 ([20]) Let f and g be two fuzzy subsets of a po-Γ-semigroup S. Then

$$(f \circ g)(x) = \begin{cases} \sup\limits_{x \leq y\gamma z} \{\min\{f(y), g(z)\}\} \text{ if there exist } y, z \in S, \gamma \in \Gamma \text{ with } x \leq y\gamma z, \\ 0 \text{ otherwise.} \end{cases}$$

Definition 2.15 ([20]) A nonempty fuzzy subset f of a po-Γ-semigroup S is called a fuzzy left (right) ideal of S if

(1) $f(x\alpha y) \geq f(y)$ $(f(x\alpha y) \geq f(x))$, for all $x, y \in S$, $\alpha \in \Gamma$,
(2) $b \leq a \Rightarrow f(b) \geq f(a)$, for all $a, b \in S$.

f is called a fuzzy ideal if f is both fuzzy left ideal and fuzzy right ideal.

3 Some Results of Fuzzy Points in Po-Γ-Semigroups

Definition 3.1 ([15]) Let S be a po-Γ-semigroup of S. Let $a \in S$ and $t \in (0, 1]$. We define a fuzzy subset a_t of S as follows:

$$a_t(x) = \begin{cases} t \text{ if } x \leq a, \\ 0 \text{ otherwise} \end{cases}$$

for all $x \in S$. We call a_t a fuzzy point or fuzzy singleton of S.

Theorem 3.2 ([15]) *Let a_t and b_r be two fuzzy points of a po-Γ-semigroup S. Then*

$$a_t \circ b_r = \bigcup_{\gamma \in \Gamma} (a\gamma b)_{t \wedge r}.$$

Remark 3.3 For any fuzzy subset f of a po-Γ-semigroup S, $f = \underset{a_t \subseteq f}{\cup} a_t$.

The following lemma follows easily.

Lemma 3.4 *Let S be a po-Γ-semigroup, f, g, and h be fuzzy subsets of S. Then $f \circ (g \cup h) = (f \circ g) \cup (f \circ h)$.*

Definition 3.5 Let S be a po-Γ-semigroup and a_t be a fuzzy point of S. Then the fuzzy ideal generated by a_t denoted by $< a_t >$, is defined to be the smallest fuzzy ideal containing a_t in S.

Proposition 3.6 *Let S be a po-Γ-semigroup and a_t be a fuzzy point of S. Then the fuzzy ideal $< a_t >$ generated by a_t is given by*

$$< a_t > (x) = \begin{cases} t, & \textit{if } x \in < a >, \\ 0, & \textit{otherwise}, \end{cases}$$

for any $x \in S$, where $< a >$ is the ideal of S generated by a.

Proof Let us consider a fuzzy subset g of S defined by

$$g(x) = \begin{cases} t, & \text{if } x \in < a >, \\ 0, & \text{otherwise}, \end{cases}$$

for any $x \in S$, where $< a >$ is the ideal of S generated by a. Let $x, y \in S$ and $\gamma \in \Gamma$. If $x, y \in < a >$, then $x\gamma y \in < a >$. So $g(x\gamma y) = t = g(x) = g(y)$. Again if $x, y \notin < a >$ but $x\gamma y \in < a >$, then $g(x\gamma y) = t \geq 0 = g(x) = g(y)$. If $x, y \notin < a >$ and $x\gamma y \notin < a >$, then $g(x\gamma y) = 0 = g(x) = g(y)$. Again if $x \in < a >$ and $y \notin < a >$, then $x\gamma y \in < a >$. So $g(x\gamma y) = t = g(x) \geq 0 = g(y)$. Let $x \leq y$ in S. If $y \in < a >$, then $x \in < a >$ whence $g(x) = t = g(y)$. Again if $y \notin < a >$, then $g(x) \geq 0 = g(y)$. So g is a fuzzy ideal.

Let f be a fuzzy ideal of S such that $a_t \subseteq f$. Then $f(a) \geq a_t(a) = t$. Now let $z \in < a >= (\{a\} \cup a\Gamma S \cup S\Gamma a \cup S\Gamma a\Gamma S]$. If $z \leq a$, then $f(z) \geq f(a) \geq t$. If $z \leq a\alpha x$ for some $x \in S$ and $\alpha \in \Gamma$, then $f(z) \geq f(a\alpha x) \geq f(a) \geq t$ (*cf.* Definition 2.15). Similarly, we can show that $f(z) \geq t$ if $z \leq y\beta a$ or $z \leq x\alpha a\beta y$ for some $x, y \in S$ and $\alpha, \beta \in \Gamma$. It follows that $g \subseteq f$. Since $g(x) \geq a_t(x)$, for all $x \in S$, g contains a_t. This completes the proof. □

Remark 3.7 From the above result, we notice that $< a_t >= tC_{<a>}$ where $C_{<a>}$ is the characteristic function of $< a >$.

Proposition 3.8 *Let S be a po-Γ-semigroup and a_t be a fuzzy point of S. Then*

$$S \circ a_t \circ S(x) = \begin{cases} t, & \textit{if } x \in (S\Gamma a\Gamma S], \\ 0, & \textit{otherwise}, \end{cases}$$

for all $x \in S$. Moreover, $S \circ a_t \circ S$ is a fuzzy ideal of S.

Proof Let $x \in S$. If $x \not\leq w\alpha z\beta y$ for any $w, z, y \in S$ and $\alpha, \beta \in \Gamma$, then $x \notin (S\Gamma a\Gamma S]$ and $S \circ a_t \circ S(x) = 0$. Now let $x \leq w\alpha z\beta y$ for some $w, z, y \in S$ and $\alpha, \beta \in \Gamma$. Then

$$\begin{aligned} S \circ a_t \circ S(x) &= \bigvee_{x \leq p\gamma q} \{S \circ a_t(p) \wedge S(q)\} \\ &= \bigvee_{x \leq p\gamma q} \{S \circ a_t(p)\} \\ &= \bigvee_{x \leq s\delta r\gamma q} \{S(s) \wedge a_t(r)\} \\ &= \bigvee_{x \leq s\delta r\gamma q} a_t(r). \end{aligned}$$

If there exists one $r = a$, then $a_t(r) = t$ whence $S \circ a_t \circ S(x) = t$. Thus if $x \in (S\Gamma a\Gamma S]$, then $S \circ a_t \circ S(x) = t$, otherwise $S \circ a_t \circ S(x) = 0$.

In order to prove the last part we see $S \circ (S \circ a_t \circ S) \subseteq S \circ a_t \circ S$ and $(S \circ a_t \circ S) \circ S \subseteq S \circ a_t \circ S$. Let $x \leq y$ in S. If $x \in (S\Gamma a\Gamma S]$, then $S \circ a_t \circ S(x) = t \geq S \circ a_t \circ S(y)$. If $x \notin (S\Gamma a\Gamma S]$, then $y \notin (S\Gamma a\Gamma S]$ whence $S \circ a_t \circ S(x) = 0 = S \circ a_t \circ S(y)$. Hence $S \circ a_t \circ S$ is a fuzzy ideal of S. □

The following result is an easy consequence of the above proposition.

Corollary 3.9 *Let S be a po-Γ-semigroup and a_t be a fuzzy point of S. Then*

$$S \circ a_t(x) = \begin{cases} t, & \text{if } x \in (S\Gamma a], \\ 0, & \text{otherwise,} \end{cases}$$

and

$$a_t \circ S(x) = \begin{cases} t, & \text{if } x \in (a\Gamma S], \\ 0, & \text{otherwise,} \end{cases}$$

for all $x \in S$. Moreover, $S \circ a_t$ is a fuzzy left ideal of S and $a_t \circ S$ is a fuzzy right ideal of S.

Remark 3.10 From Proposition 3.8 and Corollary 3.9 we notice that $S \circ a_t \circ S = tC_{(S\Gamma a\Gamma S]}$, $S \circ a_t = tC_{(S\Gamma a]}$, and $a_t \circ S = tC_{(a\Gamma S]}$.

Proposition 3.11 *Let S be a po-Γ-semigroup and a_t be a fuzzy point of S. Then $< a_t >= a_t \cup a_t \circ S \cup S \circ a_t \cup S \circ a_t \circ S$.*

Proof By Proposition 3.6, for any $x \in S$,

$$< a_t > (x) = \begin{cases} t, & \text{if } x \in \ < a >, \\ 0, & \text{otherwise.} \end{cases}$$

Let $x \in S$. If $x \notin < a >$, then $< a_t > (x) = 0$. In view of Proposition 2.11, $< a > = (\{a\} \cup S\Gamma a \cup a\Gamma S \cup S\Gamma a\Gamma S] = (\{a\}] \cup (S\Gamma a] \cup (a\Gamma S] \cup (S\Gamma a\Gamma S]$. So $x \notin (S\Gamma a\Gamma S]$ whence $S \circ a_t \circ S(x) = 0$; $x \notin (S\Gamma a]$ whence $S \circ a_t(x) =$

0; $x \notin (a\Gamma S]$ whence $a_t \circ S(x) = 0$; and $x \nleq a$ whence $a_t(x) = 0$. Hence $a_t \cup a_t \circ S \cup S \circ a_t \cup S \circ a_t \circ S(x) = 0$. If $x \in <a>$, then $<a_t>(x) = t$. Again in view of Proposition 2.11, $<a> = (\{a\} \cup S\Gamma a \cup a\Gamma S \cup S\Gamma a\Gamma S] = (\{a\}] \cup (S\Gamma a] \cup (a\Gamma S] \cup (S\Gamma a\Gamma S]$. Now $x \in (S\Gamma a\Gamma S]$ whence $S \circ a_t \circ S(x) = t$; $x \in (S\Gamma a]$ whence $S \circ a_t(x) = t$; $x \in (a\Gamma S]$ whence $a_t \circ S(x) = t$; and $x \leq a$ whence $a_t(x) = t$. Hence $a_t \cup a_t \circ S \cup S \circ a_t \cup S \circ a_t \circ S(x) = t$. Consequently, $<a_t> = a_t \cup a_t \circ S \cup S \circ a_t \cup S \circ a_t \circ S$. □

We omit the proof of the following Corollary since it is similar to that of Corollary 1 of [21].

Corollary 3.12 *Let S be a po-Γ-semigroup and a_t be a fuzzy point of S. Then $<a_t>^3 \subseteq S \circ a_t \circ S$.*

Though the following proposition is easy to obtain, it is also useful for the development of the paper.

Proposition 3.13 *Let S be a po-Γ-semigroup, A and B be subset of S and C_A be the characteristic function of A. Then for any $t, r \in (0, 1]$, the following statements hold.*

(i) $tC_A \circ rC_B = (t \wedge r)C_{(A\Gamma B]}$.
(ii) $tC_A \cap tC_B = tC_{A\cap B}$.
(iii) $tC_{(A]} = \bigcup_{a\in A} a_t$.
(iv) $S \circ tC_A = tC_{(S\Gamma A]}$.
(v) *A is an ideal (right ideal, left ideal) of S if and only if tC_A is a fuzzy ideal (fuzzy right ideal, fuzzy left ideal) of S.*

4 Prime Fuzzy Ideals and Weakly Prime Fuzzy Ideals in Po-Γ-Semigroups

In this section, we deduce various properties and characterizations of prime fuzzy ideals and weakly prime fuzzy ideals of po-Γ-semigroups.

Definition 4.1 ([3]) Let S be a po-Γ-semigroup. Then an ideal $I(\neq S)$ of S is called prime if for any two ideals A and B of S, $A\Gamma B \subseteq I$ implies $A \subseteq I$ or $B \subseteq I$.

Definition 4.2 ([3]) Let S be a po-Γ-semigroup. Then an ideal $I(\neq S)$ of S is called completely prime if for any $a, b \in S$, $a\Gamma b \subseteq I$ implies $a \in I$ or $b \in I$.

Definition 4.3 Let S be a po-Γ-semigroup. Then a fuzzy ideal f of S is called prime fuzzy ideal if f is a nonconstant function and for any two fuzzy ideals g and h of S, $g \circ h \subseteq f$ implies $g \subseteq f$ or $h \subseteq f$.

Example 4.4 Let $S = \mathbb{Z}_0^-$ and $\Gamma = \mathbb{Z}_0^-$, where $\mathbb{Z}_0^-$ denotes the set of all negative integers with 0. Then S is a Γ-semigroup where $a\gamma b$ denotes the usual multiplication of integers a, γ, b where $a, b \in S$ and $\gamma \in \Gamma$. Again with respect to usual $\leq$ of $\mathbb{Z}$, S becomes a po-Γ-semigroup. Let p be a prime number. Now we define a fuzzy subset f on S by

$$f(x) = \begin{cases} 1, & \text{for } x \in (p\mathbb{Z}_0^-], \\ 0.6, & \text{otherwise.} \end{cases}$$

Then f is a prime fuzzy ideal of S.

Example 4.5 Let $S = \{a, b, c\}$. Let $\Gamma = \{\alpha, \beta\}$ be the nonempty set of binary operations on S with the following Cayley tables.

α	a	b	c	β	a	b	c
a	a	b	b	a	b	b	b
b	b	b	b	b	b	b	b
c	c	c	c	c	c	c	c

By a routine verification, we see that S is a po-Γ-semigroup where the partial orders on S and Γ are given by $c \leq b \leq a$ and $\beta \leq \alpha$, respectively. Now we define a fuzzy subset μ on S by $\mu(a) = 0.5$, $\mu(b) = 1 = \mu(c)$. It is easy to observe that μ is a prime fuzzy ideal of S.

Though the proof of the following theorem is straightforward, it also characterizes a prime fuzzy ideal.

Theorem 4.6 *Let S be a commutative po-Γ-semigroup and f be a fuzzy ideal of S. Then f is prime fuzzy ideal if and only if for any fuzzy subsets g and h of S, $g \circ h \subseteq f$ implies $g \subseteq f$ or $h \subseteq f$.*

Theorem 4.7 *Let S be a po-Γ-semigroup and I be an ideal of S. Then I is a prime ideal of S if and only if C_I, the characteristic function of I, is a prime fuzzy ideal of S.*

Proof Let I be a prime ideal of S. Then C_I is a fuzzy ideal of S (*cf.* Proposition 3.13). Now let f and g be two fuzzy ideals of S such that $f \circ g \subseteq C_I$ and $f \nsubseteq C_I$. Then there exists a fuzzy point $x_t \subseteq f$ $(t > 0)$ such that $x_t \nsubseteq C_I$. Let $y_r \subseteq g$ $(r > 0)$. Then $< x_t > \circ < y_r > \subseteq f \circ g \subseteq C_I$. Again for all $z \in S$, in view of Propositions 3.6 and 3.13, we obtain

$$< x_t > \circ < y_r > (z) = \begin{cases} t \wedge r, & \text{if } z \in (< x > \Gamma < y >], \\ 0, & \text{otherwise.} \end{cases}$$

Hence $(< x > \Gamma < y >] \subseteq I$. Using Proposition 2.11 we see that $(< x >]\Gamma(< y >] \subseteq (< x > \Gamma < y >]$ whence $(< x >]\Gamma(< y >] \subseteq I$. This together with the hypothesis implies that $(< x >] \subseteq I$ or $(< y >] \subseteq I$ (*cf.* Definition 4.1). As

$< x > \subseteq (< x >]$ and $< y > \subseteq (< y >]$, we have $< x > \subseteq I$ or $< y > \subseteq I$. Since $x_t \nsubseteq C_I$, $t = x_t(x) > C_I(x)$. So $C_I(x) = 0$ whence $x \notin I$. Hence $< x > \nsubseteq I$. Consequently, $< y > \subseteq I$. Then $y_r \subseteq C_I$ and so $g \subseteq C_I$. Hence C_I is a prime fuzzy ideal of S.

Conversely, suppose C_I is a prime fuzzy ideal of S. Then C_I is a fuzzy ideal of S which together with Proposition 3.13 implies that I is an ideal of S. Let A and B be two fuzzy ideals of S such that $A\Gamma B \subseteq I$. Then $(A\Gamma B] \subseteq I$. Again C_A, and C_B are fuzzy ideals of S and $C_A \circ C_B = C_{(A\Gamma B]} \subseteq C_I$ (*cf.* Proposition 3.13). So by hypothesis, $C_A \subseteq C_I$ or $C_B \subseteq C_I$. Hence $A \subseteq I$ or $B \subseteq I$. Consequently, I is a prime ideal of S. □

Proposition 4.8 *Let S be a po-Γ-semigroup and f be a prime fuzzy ideal of S. Then* $|Imf| = 2$.

Proof By Definition 4.3, f is a nonconstant fuzzy ideal. So $|Imf| \geq 2$. Suppose $|Imf| > 2$. Then there exist $x, y, z \in S$ such that $f(x), f(y), f(z)$ are distinct. Let us assume, without loss of generality, $f(x) < f(y) < f(z)$. Then there exist $r, t \in (0, 1)$ such that $f(x) < r < f(y) < t < f(z) \cdots (1)$. Then for all $u \in S$,

$$< x_r > \circ < y_t > (u) = \begin{cases} r \wedge t, & \text{if } u \in (< x > \Gamma < y >], \\ 0, & \text{otherwise.} \end{cases}$$

Let $u \in (< x > \Gamma < y >]$. Then $u \leq a\gamma b$ where $a \in < x >, b \in < y >$ and $\gamma \in \Gamma$. Since f is a fuzzy ideal of S, $f(u) \geq f(a\gamma b) \geq f(x) \vee f(y) > r \wedge t$. Therefore $< x_r > \circ < y_t > \subseteq f$ which, by Definition 4.3, implies that $< x_r > \subseteq f$ or $< y_t > \subseteq f$. Suppose $< x_r > \subseteq f$. Then $f(x) \geq < x_r > (x) = r$ which contradicts (1). Similarly, $< y_t > \subseteq f$ contradicts (1). Hence $|Imf| = 2$. □

Theorem 4.9 *Let S be a po-Γ-semigroup and f be a prime fuzzy ideal of S. Then there exists* $x_0 \in S$ *such that* $f(x_0) = 1$.

Proof By Proposition 4.8, we have $|Imf| = 2$. Suppose $Imf = \{t, s\}$ such that $t < s$. Let if possible $f(x) < 1$, for all $x \in S$. Then $t < s < 1$. Let $f(x) = t$ and $f(y) = s$ for some $x, y \in S$. Then $f(x) = t < s = f(y) < 1$. Now we choose $t_1, t_2 \in (0, 1)$ such that $f(x) < t_1 < f(y) < t_2 < 1$. Then by the similar argument as applied in the proof of Proposition 4.8, we obtain $< x_{t_1} > \circ < y_{t_2} > \subseteq f$. Since f is a prime fuzzy ideal of S, $< x_{t_1} > \subseteq f$ or $< y_{t_2} > \subseteq f$ whence $f(x) \geq t_1$ or $f(y) \geq t_2$. This contradicts the choices of t_1 and t_2. Hence there exists an $x_0 \in S$ such that $f(x_0) = 1$. □

Theorem 4.10 *Let S be a po-Γ-semigroup and f be a prime fuzzy ideal of S. Then each level subset* f_t $(\neq S)$, $t \in (0, 1]$, *if nonempty, is a prime ideal of S.*

Proof Since f is a fuzzy ideal, each level subset f_t, $t \in (0, 1]$, if nonempty, is an ideal of S (*cf.* Theorem 3.5 [20]). Let $t \in (0, 1]$ be such that f_t $(\neq S)$ is nonempty. Now let I, J be two ideals of S such that $I\Gamma J \subseteq f_t$. Since f_t is an ideal of S, $(I\Gamma J] \subseteq f_t$. Then $tC_{(I\Gamma J]} \subseteq f$. By Proposition 3.13(v), $g := tC_I$ and $h := tC_J$

are fuzzy ideals of S. Since $g \circ h = tC_I \circ tC_J = tC_{(I\Gamma J]}$ (*cf.* Proposition 3.13(i)), $g \circ h \subseteq f$. Since f is a prime fuzzy ideal, $g \subseteq f$ or $h \subseteq f$. Hence either $tC_I \subseteq f$ or $tC_J \subseteq f$ whence we obtain $I \subseteq f_t$ or $J \subseteq f_t$. Hence f_t is a prime ideal of S. □

As a consequence of Theorems 4.9 and 4.10, we obtain the following result.

Corollary 4.11 *If f is a prime fuzzy ideal of a po-Γ-semigroup S, then the level subset f_1 is a prime ideal of S.*

Remark 4.12 The converse of Theorem 4.10 is not true which is illustrated in the following example.

Example 4.13 Let S be a po-Γ-semigroup and A be a prime ideal of S. Let

$$f(x) = \begin{cases} t, & \text{if } x \in A, \\ 0, & \text{otherwise.} \end{cases}$$

Then f is a fuzzy ideal of S. Here $f_{t_1} = A$, where $0 < t_1 \leq t$. Hence each of nonempty level subsets of f is a prime ideal of S. But if $0 < t < 1$, then f is not a prime fuzzy ideal of S (*cf.* Theorem 4.9).

Lemma 4.14 *Let S be a po-Γ-semigroup. Then a fuzzy subset f of S satisfying (i) and (ii),*

(i) $|Imf| = 2$,
(ii) *f_1 is an ideal of S,*

is a fuzzy ideal of S.

The following result also characterizes a prime fuzzy ideal of a po-Γ-semigroup.

Theorem 4.15 *Let S be a po-Γ-semigroup. Then a fuzzy subset f of S is a prime fuzzy ideal of S if and only if f satisfies the following conditions:*

(i) $|Imf| = 2$.
(ii) *f_1 is a prime ideal of S.*

Proof The direct implication follows easily from Proposition 4.8, Theorem 4.9 and Corollary 4.11.

To prove the converse, we first observe that f is a fuzzy ideal of S (*cf.* Lemma 4.14). Then let g and h be two fuzzy ideals of S such that $g \circ h \subseteq f$. If $g \nsubseteq f$ and $h \nsubseteq f$, then there exist $x, y \in S$ such that $g(x) > f(x)$ and $h(y) > f(y)$. Thus $x, y \notin f_1$. We claim that $x\Gamma S\Gamma y \nsubseteq f_1$. To establish the claim we suppose the contrary. Then $(S\Gamma x\Gamma S\Gamma S\Gamma y\Gamma S] \subseteq f_1$ (as f_1 is an ideal) and so $(S\Gamma x\Gamma S]\Gamma(S\Gamma y\Gamma S] \subseteq (S\Gamma x\Gamma S\Gamma S\Gamma y\Gamma S] \subseteq f_1$ (*cf.* Proposition 2.11) whence $(S\Gamma x\Gamma S] \subseteq f_1$ or $(S\Gamma y\Gamma S] \subseteq f_1$ (as f_1 is a prime ideal). Let us assume without loss of generality $(S\Gamma x\Gamma S] \subseteq f_1$. Then $< x >^3 \subseteq (S\Gamma x\Gamma S] \subseteq f_1$ which implies that $x \in < x > \subseteq f_1$, which is a contradiction. Hence $x\Gamma S\Gamma y \nsubseteq f_1$. Then there exist $s \in S$, $\alpha, \beta \in \Gamma$ such that $x\alpha s\beta y \notin f_1$ which means $f(x\alpha s\beta y) < 1$. So

$f(x\alpha s\beta y) = t = f(x) = f(y)$, where $Imf = \{t, 1\}$. But by using Definitions 2.14 and 2.15, we obtain

$$\begin{aligned}(g \circ h)(x\alpha s\beta y) &\geq g(x) \wedge h(s\beta y)\\ &\geq g(x) \wedge h(y)\\ &> f(x) \wedge f(y)\\ &= t.\end{aligned}$$

Hence $g \circ h \nsubseteq f$ which is a contradiction. Hence f is a prime fuzzy ideal of S. □

Corollary 4.16 *Let S be a po-Γ-semigroup and f be a prime fuzzy ideal of S. Then there exists a prime fuzzy ideal g of S such that f is properly contained in g.*

Proof By Theorem 4.15, there exists $x_0 \in S$ such that $f(x_0) = 1$ and $Im(f) = \{t, 1\}$ for some $t \in [0, 1)$. Let g be a fuzzy subset of S defined by $g(x) = 1$, if $x \in f_1$ and $g(x) = r$, if $x \notin f_1$, where $t < r < 1$. Then by Theorem 4.15, g is a prime fuzzy ideal and $f \subsetneqq g$. □

In Theorem 4.10 we have shown that every nonempty level subset of a prime fuzzy ideal is a prime ideal. But Example 4.13 shows that the converse need not be true. In order to make the level subset criterion to hold, a new type of fuzzy primeness in ideals of a po-Γ-semigroup can be defined what is called weakly prime fuzzy ideal.

Definition 4.17 Let S be a po-Γ-semigroup. A nonconstant fuzzy ideal f of S is called a weakly prime fuzzy ideal of S if for all ideals A and B of S and for all $t \in (0, 1]$, $tC_A \circ tC_B \subseteq f$ implies $tC_A \subseteq f$ or $tC_B \subseteq f$.

Theorem 4.18 *Let S be a po-Γ-semigroup and f be a fuzzy ideal of S. Then f is a weakly prime fuzzy ideal of S if and only if each level subset f_t ($\neq S$), $t \in (0, 1]$, is a prime ideal of S for $f_t \neq \varnothing$.*

Proof Let f be a weakly prime fuzzy ideal of S and $t \in (0, 1]$ such that $f_t \neq \varnothing$ and $f_t \neq S$. Then f is a fuzzy ideal of S. So f_t is an ideal of S (*cf.* Theorem 3.5 [20]). Let A and B be ideals of S with $A\Gamma B \subseteq f_t$. Then f_t being an ideal of S, $(A\Gamma B] \subseteq f_t$. Therefore, $tC_{(A\Gamma B]} \subseteq f$ which means $tC_A \circ tC_B \subseteq f$ (*cf.* Proposition 3.13). Hence by hypothesis, $tC_A \subseteq f$ or $tC_B \subseteq f$ (*cf.* Definition 4.17). Hence either $A \subseteq f_t$ or $B \subseteq f_t$. Consequently, f_t is a prime ideal of S.

Conversely, suppose each f_t ($\neq S$) is a prime ideal of S, for all $t \in (0, 1]$ with $f_t \neq \varnothing$. Let A and B be ideals of S such that $tC_A \circ tC_B \subseteq f$ where $t \in (0, 1]$. Then $tC_{(A\Gamma B]} \subseteq f$ (*cf.* Proposition 3.13) whence $(A\Gamma B] \subseteq f_t$. Now by Proposition 2.11, $(A]\Gamma(B] \subseteq (A\Gamma B] \subseteq f_t$ whence $A\Gamma B \subseteq f_t$ (as A, B are ideals). Hence by hypothesis either $A \subseteq f_t$ or $B \subseteq f_t$ whence $tC_A \subseteq f$ or $tC_B \subseteq f$. Hence f is a weakly prime fuzzy ideal of S. □

As an easy consequence of Theorems 4.10 and 4.18, we obtain the following corollary.

Corollary 4.19 *In a po-Γ-semigroup S, every prime fuzzy ideal is a weakly prime fuzzy ideal.*

That the converse of the above corollary is not always true is illustrated in the following examples.

Example 4.20 The fuzzy ideal of Example 4.13 is a weakly prime fuzzy ideal (*cf.* Theorem 4.18) but not a prime fuzzy ideal.

Example 4.21 Let $S = \{0, a, b, c\}$. Let $\Gamma = \{\alpha, \beta, \gamma\}$ be the nonempty set of binary operations on S with the following Cayley tables.

α	0	a	b	c	β	0	a	b	c	γ	0	a	b	c
0	0	0	0	0	0	0	0	0	0	0	0	0	0	0
a	0	0	0	b	a	0	0	0	b	a	0	0	0	0
b	0	0	0	b	b	0	0	0	b	b	0	0	0	0
c	b	b	b	c	c	b	b	b	b	c	0	0	0	0

By a routine but tedious verification, we see that S is a po-Γ-semigroup where the partial orders on S and Γ are given by $0 \leq a \leq b \leq c$ and $\gamma \leq \beta \leq \alpha$, respectively. Now we define a fuzzy subset f on S by $f(0) = f(a) = 0.8$, $f(b) = 0.3$, and $f(c) = 0$. It can be checked that f is a weakly prime fuzzy ideal of S. But f is not a prime fuzzy ideal of S (*cf.* Theorem 4.15).

Remark 4.22 The above corollary and the example together shows that the notion of weakly prime fuzzy ideal generalizes the notion of prime fuzzy ideal.

The following theorem characterizes weakly prime fuzzy ideals of po-Γ-semigroups.

Theorem 4.23 *Let S be a po-Γ-semigroup and f be a fuzzy ideal of S. Then the following are equivalent.*

(i) *f is a weakly prime fuzzy ideal of S.*
(ii) *For any $x, y \in S$ and $r \in (0, 1]$, if $x_r \circ S \circ y_r \subseteq f$, then $x_r \subseteq f$ or $y_r \subseteq f$.*
(iii) *For any $x, y \in S$ and $r \in (0, 1]$, if $< x_r > \circ < y_r > \subseteq f$, then $x_r \subseteq f$ or $y_r \subseteq f$.*
(iv) *If A and B are right ideals of S such that $tC_A \circ tC_B \subseteq f$, then $tC_A \subseteq f$ or $tC_B \subseteq f$.*
(v) *If A and B are left ideals of S such that $tC_A \circ tC_B \subseteq f$, then $tC_A \subseteq f$ or $tC_B \subseteq f$.*
(vi) *If A is a right ideal of S and B is a left ideal of S such that $tC_A \circ tC_B \subseteq f$, then $tC_A \subseteq f$ or $tC_B \subseteq f$.*

Proof (i) $\Rightarrow$ (ii).

Let f be a weakly prime fuzzy ideal of S. Let $x, y \in S$ and $r \in (0, 1]$ be such that $x_r \circ S \circ y_r \subseteq f$. Then by Proposition 3.8, $rC_{(S\Gamma x\Gamma S]} \circ rC_{(S\Gamma y\Gamma S]} =$

$(S \circ x_r \circ S) \circ (S \circ y_r \circ S) \subseteq S \circ (x_r \circ S \circ y_r) \circ S \subseteq S \circ f \circ S \subseteq f$. Hence by hypothesis, $rC_{(S\Gamma x\Gamma S]} \subseteq f$ or $rC_{(S\Gamma y\Gamma S]} \subseteq f$ whence $S \circ x_r \circ S \subseteq f$ or $S \circ y_r \circ S \subseteq f$. If $S \circ x_r \circ S \subseteq f$, then $< x_r >^3 \subseteq f$ (*cf.* Corollary 3.12). Hence $(rC_{<x>})^3 \subseteq f$. Since f is weakly prime fuzzy ideal, this implies that $< x_r > \subseteq f$ whence $x_r \subseteq f$. Similarly, if $S \circ y_r \circ S \subseteq f$, then $y_r \subseteq f$.
(ii) $\Rightarrow$ (iii).

Let $x, y \in S$ and $r \in (0, 1]$ be such that $< x_r > \circ < y_r > \subseteq f$. Then since $x_r \circ S \subseteq < x_r >$ and $y_r \subseteq < y_r >$, $x_r \circ S \circ y_r \subseteq f$. Hence by (ii), $x_r \subseteq f$ or $y_r \subseteq f$.
(iii) $\Rightarrow$ (iv).

Let A, B be two right ideals of S such that $tC_A \circ tC_B \subseteq f$ and $tC_A \nsubseteq f$. Then there exists $a \in A$ such that $a_t \nsubseteq f$. Now for any $b \in B$, by Proposition 3.13 and hypothesis, we obtain $< a_t > \circ < b_t > = tC_{<a>} \circ tC_{<b>} = tC_{(<a>\Gamma<b>]}$ $\subseteq tC_{((A\Gamma B)\cup(S\Gamma A\Gamma B))]} = tC_{(A\Gamma B]\cup(S\Gamma A\Gamma B]} = tC_{(A\Gamma B]} \cup tC_{(S\Gamma A\Gamma B]} = tC_{(A\Gamma B]} \cup tC_{(S\Gamma(A]\Gamma(B]]} \subseteq tC_{(A\Gamma B]} \cup tC_{(S\Gamma(A\Gamma B]]} = (tC_A \circ tC_B) \cup (S \circ tC_{(A\Gamma B]}) = (tC_A \circ tC_B) \cup (S \circ tC_A \circ tC_B)$ (*cf.* Proposition 3.13) $\subseteq f \cup (S \circ f) \subseteq f$. Hence by (iii), $b_t \subseteq f$. Consequently, $tC_B \subseteq f$.
(iii) $\Rightarrow$ (vi).

Let A be a right ideal and B be a left ideal of S such that $tC_A \circ tC_B \subseteq f$ and $tC_A \nsubseteq f$. Then there exists $a \in A$ such that $a_t \nsubseteq f$. Now for any $b \in B$, $< a_t > \circ < b_t > = tC_{<a>} \circ tC_{<b>} = tC_{(<a>\Gamma<b>]} \subseteq tC_{((A\Gamma B)\cup(A\Gamma B\Gamma S)\cup(S\Gamma A\Gamma B)\cup(S\Gamma A\Gamma B\Gamma S))]}$ $= tC_{(A\Gamma B]\cup(A\Gamma B\Gamma S]\cup(S\Gamma A\Gamma B]\cup(S\Gamma A\Gamma B\Gamma S]} = tC_{(A\Gamma B]} \cup tC_{(A\Gamma B\Gamma S]} \cup tC_{(S\Gamma A\Gamma B]} \cup tC_{(S\Gamma A\Gamma B\Gamma S]} = tC_{(A\Gamma B]} \cup tC_{((A]\Gamma(B]\Gamma S]} \cup tC_{(S\Gamma(A]\Gamma(B]]} \cup tC_{(S\Gamma(A]\Gamma(B]\Gamma S]} \subseteq tC_{(A\Gamma B]} \cup tC_{((A\Gamma B]\Gamma S]} \cup tC_{(S\Gamma(A\Gamma B]]} \cup tC_{(S\Gamma(A\Gamma B]\Gamma S]} = (tC_A \circ tC_B) \cup (tC_A \circ tC_B \circ S) \cup (S \circ tC_A \circ tC_B) \cup (S \circ tC_A \circ tC_B \circ S) \subseteq f \cup (f \circ S) \cup (S \circ f) \cup (S \circ f \circ S)$ $\subseteq f$. Hence by (iii), $b_t \subseteq f$. Consequently, $tC_B \subseteq f$.

(iv) $\Rightarrow$ (i), (v) $\Rightarrow$ (i), (vi) $\Rightarrow$ (i) are obvious and (iii) $\Rightarrow$ (v) is similar to (iii) $\Rightarrow$ (iv). □

5 Completely Prime and Weakly Completely Prime Fuzzy Ideals in Po-Γ-Semigroups

In this section, the notion of completely prime ideals of po-Γ-semigroups has been generalized in fuzzy setting.

Definition 5.1 Let S be a po-Γ-semigroup. A nonconstant fuzzy ideal f of S is called a completely prime fuzzy ideal if for any two fuzzy points x_t, y_r of S $(t, r \in (0, 1])$, $x_t \circ y_r \subseteq f$ implies that $x_t \subseteq f$ or $y_r \subseteq f$.

Definition 5.2 Let S be a po-Γ-semigroup. A nonconstant fuzzy ideal f of S is called a weakly completely prime fuzzy ideal if for any fuzzy points x_t, y_t of S $(t \in (0, 1])$, $x_t \circ y_t \subseteq f$ implies that $x_t \subseteq f$ or $y_t \subseteq f$.

Theorem 5.3 *Let S be a po-Γ-semigroup and f be a fuzzy ideal of S. Then f is completely prime fuzzy ideal if and only if for any fuzzy subsets f and g of S, $g \circ h \subseteq f$ implies $g \subseteq f$ or $h \subseteq f$.*

Proof Let f be a completely prime fuzzy ideal and g, h be two fuzzy subsets such that $g \circ h \subseteq f$ and $g \nsubseteq f$. Then there exists an $x_t \subseteq g$ such that $x_t \nsubseteq f$. Since $g \circ h \subseteq f$, $x_t \circ y_r \subseteq f$, for all $y_r \subseteq h$. So $y_r \subseteq f$, for all $y_r \subseteq h$. Hence $h \subseteq f$.

The converse follows easily. □

Definitions 5.1, 5.2 and Theorems 5.3 and 4.6 together give rise to the following result.

Corollary 5.4 *Let S be a po-Γ-semigroup and f be a completely prime fuzzy ideal of S. Then f is a prime fuzzy ideal and a weakly completely prime fuzzy ideal of S. Further if S is commutative, then f is a prime fuzzy ideal if and only if f is a completely prime fuzzy ideal.*

The following theorem characterizes a completely prime fuzzy ideal.

Theorem 5.5 *Let S be a po-Γ-semigroup and f be a fuzzy subset of S. Then f is a completely prime fuzzy ideal of S if and only if f satisfies the following conditions:*

(1) $|Imf| = 2$.
(2) *f_1 is a completely prime ideal of S.*

Proof Let f be a completely prime fuzzy ideal of S. Then by Corollary 5.4, f is a prime fuzzy ideal of S. So by Theorem 4.15, f_1 is a prime ideal and $|Imf| = 2$. Let $x, y \in S$ such that $x\Gamma y \in f_1$. Then $f(x\gamma y) = 1$, for all $\gamma \in \Gamma$. So the fuzzy point $(x\gamma y)_1 \subseteq f$, for all $\gamma \in \Gamma$ whence $\underset{\gamma\in\Gamma}{\cup}(x\gamma y)_1 \subseteq f$. Therefore $x_1 \circ y_1 \subseteq f$ (*cf.* Theorem 3.2). Since f is a completely prime fuzzy ideal, $x_1 \subseteq f$ or $y_1 \subseteq f$ whence $x \in f_1$ or $y \in f_1$. Hence f_1 is a completely prime ideal of S.

Conversely, suppose the given conditions hold, *i.e.*, $Im(f) = \{t, 1\}$ $(t < 1)$ and f_1 is a completely prime ideal. Then f_1 is a prime ideal of S. So by Theorem 4.15, f is a prime fuzzy ideal. Hence f is a nonconstant fuzzy ideal of S. Let x_r and y_s $(r, s > 0)$ be two fuzzy points of S such that $x_r \circ y_s \subseteq f$. If possible let $x_r \nsubseteq f$ and $y_s \nsubseteq f$, then $f(x) < r$ and $f(y) < s$. So $f(x) = f(y) = t$. Thus $x, y \notin f_1$, which implies $x\Gamma y \nsubseteq f_1$ as f_1 is completely prime. So there exists $\gamma \in \Gamma$ such that $x\gamma y \notin f_1$ whence $f(x\gamma y) = t$. Now $(x_r \circ y_s)(x\gamma y) = (\underset{\beta\in\Gamma}{\cup}(x\beta y)_{r\wedge s})(x\gamma y) = r \wedge s > t = f(x\gamma y)$. This is a contradiction to $x_r \circ y_s \subseteq f$. Hence f is a completely prime fuzzy ideal of S. □

The following theorem characterizes a weakly completely prime fuzzy ideal.

Theorem 5.6 *Let S be a po-Γ-semigroup and f be a fuzzy ideal of S. Then f is weakly completely prime fuzzy ideal if and only if $\inf_{\gamma\in\Gamma} f(x\gamma y) = \max\{f(x), f(y)\}$, for all $x, y \in S$.*

Proof Let f be a weakly completely prime fuzzy ideal and $x, y \in S$. Since f is a fuzzy ideal, $f(x\gamma y) \geq \max\{f(x), f(y)\}$, for all $\gamma \in \Gamma$. So $\inf_{\gamma\in\Gamma} f(x\gamma y) \geq \max\{f(x), f(y)\}$. Now let $\inf_{\gamma\in\Gamma} f(x\gamma y) = t$ where $t \in [0, 1]$. If $t = 0$, then $\inf_{\gamma\in\Gamma} f(x\gamma y) \leq \max\{f(x), f(y)\}$. Otherwise, $f(x\gamma y) \geq t$, for all $\gamma \in \Gamma$, *i.e.*, $(x\gamma y)_t \subseteq f$, for all $\gamma \in \Gamma$. So $\bigcup_{\gamma\in\Gamma}(x\gamma y)_t \subseteq f$ which means $x_t \circ y_t \subseteq f$ (*cf.* Theorem 3.2). Since f is a weakly completely prime fuzzy ideal, $x_t \subseteq f$ or $y_t \subseteq f$, *i.e.*, $f(x) \geq t$ or $f(y) \geq t$. So $\max\{f(x), f(y)\} \geq t = \inf_{\gamma\in\Gamma} f(x\gamma y)$. Hence $\inf_{\gamma\in\Gamma} f(x\gamma y) = \max\{f(x), f(y)\}$.

Conversely, suppose the condition holds, let x_t and y_t be two fuzzy points of S such that $x_t \circ y_t \subseteq f$ where $t \in (0, 1]$. Then $\bigcup_{\gamma\in\Gamma}(x\gamma y)_t \subseteq f$ (*cf.* Theorem 3.2), *i.e.*, $(x\gamma y)_t \subseteq f$, for all $\gamma \in \Gamma$, *i.e.*, $f(x\gamma y) \geq t$, for all $\gamma \in \Gamma$ which implies $\inf_{\gamma\in\Gamma} f(x\gamma y) \geq t$. So by the hypothesis $\max\{f(x), f(y)\} \geq t$. Then $f(x) \geq t$ or $f(y) \geq t$, *i.e.*, $x_t \subseteq f$ or $y_t \subseteq f$. Hence f is a weakly completely prime fuzzy ideal of S. □

Theorem 5.7 *Let S be a po-Γ-semigroup and f be a fuzzy ideal of S. Then f is weakly completely prime fuzzy ideal if and only if each f_t, $t \in (0, 1]$, is a completely prime ideal of S for $f_t \neq \emptyset$.*

Proof Let f be a weakly completely prime fuzzy ideal of S, $x, y \in S$ and $t \in (0, 1]$ such that $f_t \neq \emptyset$. Let $x\Gamma y \subseteq f_t$. Then $f(x\gamma y) \geq t$, for all $\gamma \in \Gamma$ which means $\inf_{\gamma\in\Gamma} f(x\gamma y) \geq t$. So $\max\{f(x), f(y)\} \geq t$ (*cf.* Theorem 5.6) which implies $f(x) \geq t$ or $f(y) \geq t$, *i.e.*, $x \in f_t$ or $y \in f_t$. Hence f_t is a completely prime ideal of S.

Conversely, suppose each f_t, $t \in (0, 1]$, is a completely prime ideal of S for $f_t \neq \emptyset$. Let $x, y \in S$. Since f is a fuzzy ideal, $f(x\gamma y) \geq \max\{f(x), f(y)\}$, for all $\gamma \in \Gamma$. So $\inf_{\gamma\in\Gamma} f(x\gamma y) \geq \max\{f(x), f(y)\}$. Now let $\inf_{\gamma\in\Gamma} f(x\gamma y) = t$ where $t \in [0, 1]$. If $t = 0$, then $\inf_{\gamma\in\Gamma} f(x\gamma y) \leq \max\{f(x), f(y)\}$. Otherwise, $f(x\gamma y) \geq t$, for all $\gamma \in \Gamma$, *i.e.*, $x\gamma y \in f_t$, for all $\gamma \in \Gamma$, *i.e.*, $x\Gamma y \in f_t$. Since f_t is a completely prime fuzzy ideal, $x \in f_t$ or $y \in f_t$, *i.e.*, $f(x) \geq t$ or $f(y) \geq t$. So $\max\{f(x), f(y)\} \geq t = \inf_{\gamma\in\Gamma} f(x\gamma y)$. Hence $\inf_{\gamma\in\Gamma} f(x\gamma y) = \max\{f(x), f(y)\}$ whence f is a weakly completely prime fuzzy ideal (*cf.* Theorem 5.6). □

By Theorems 4.18 and 5.7 we have the following result.

Corollary 5.8 *Let S be a po-Γ-semigroup and f be a weakly completely prime fuzzy ideal of S. Then f is a weakly prime fuzzy ideal of S.*

Remark 5.9 Since in a both-sided po-Γ-semigroup the notions of prime ideals and completely prime ideals coincide (*cf.* Theorem 2.9 [18]), in view of Theorems 4.18 and 5.7 the notions of weakly completely prime fuzzy ideals and weakly prime fuzzy ideals coincide. Hence the above corollary is meaningless in a both-sided po-Γ-semigroup.

Remark 5.10 The proofs of results on completely prime and weakly completely prime fuzzy ideals in po-Γ-semigroups indicate that these are also true in Γ-semigroups without partial order.

To conclude this section, we give the following interrelations among various fuzzy primeness studied in this paper.

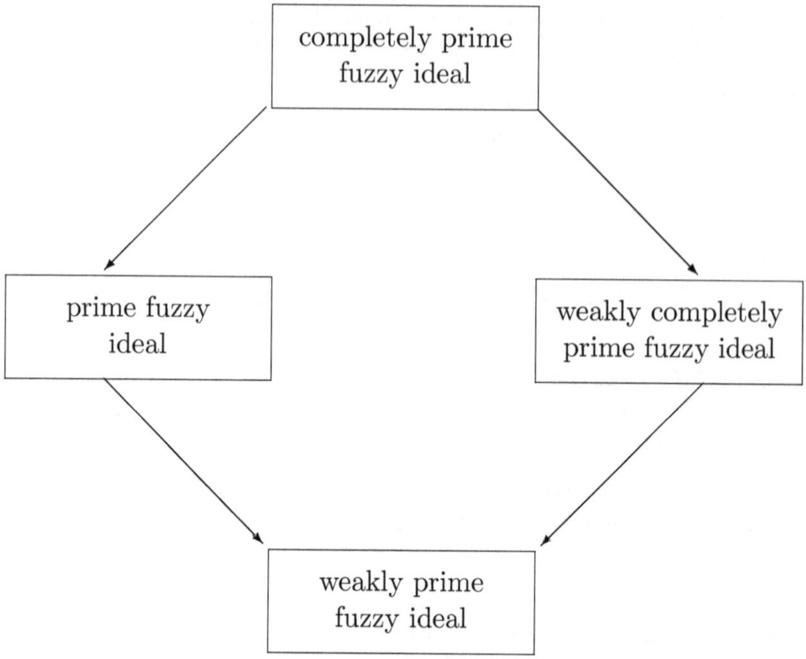

6 Concluding Remark

Theorem 4.23 is analogous to Theorem 3.4 [6]. The said theorem of [6] plays an important role in radical theory of Γ-semigroups. So Theorem 4.23 may help to study radical theory in po-Γ-semigroups via fuzzy subsets. This possibility of study of radical theory in po-Γ-semigroups is also indicated in the work of radical theory in po-semigroups via weakly prime fuzzy ideals by Xie and Tang [27].

References

1. Chattopadhyay, S.: Right inverse Γ-semigroups. Bull. Calcutta Math. Soc. **93**, 435–442 (2001)
2. Chinram, R.: On quasi-Γ-ideals in Γ-semigroups. ScienceAsia **32**, 351–353 (2006)

3. Dheena, P., Elavarasan, B.: Right chain po-Γ-semigroups. Bulletin of the Institute of Mathematics Academia Sinica (New Series) **3**(3), 407–415 (2008)
4. Dubois, D., Prade, H.: Fuzzy Sets and Systems: Theory and Applications. Academic Press, New York (1980)
5. Dutta, T.K., Adhikari, N.C.: On Γ-semigroup with the right and left unities. Soochow J. Math. **19**(4), 461–474 (1993)
6. Dutta, T.K., Adhikari, N.C.: On prime radical of Γ-semigroup. Bull. Calcutta Math. Soc. **86**(5), 437–444 (1994)
7. Dutta, T.K., Adhikari, N.C.: On partially ordered Γ-semigroup. South East Asian Bull. Math. **28**(6), 1021–1028 (2004)
8. Dutta, T.K., Chattopadhyay, S.: On uniformly strongly prime Γ-semigroup (2). Int. J. Algebra, Number Theory Appl. **1**(1), 35–42 (2009)
9. Kehayopulu, N., Tsingelis, M.: Fuzzy sets in ordered groupoids. Semigroup Forum **65**, 128–132 (2002)
10. Kehayopulu, N.: Weakly prime and prime fuzzy ideals in ordered semigroups. Lobachevskii J. Math. **27**, 31–40 (2007)
11. Kehayopulu, N.: On prime, weakly prime ideals in po-Γ-semigroups. Lobachevskii J. Math. **30**, 257–262 (2009)
12. Kuroki, N.: Fuzzy semiprime quasi ideals in semigroups. Inf. Sci. **75**(3), 201–211 (1993)
13. Kwon, Y.I., Lee, S.K.: On the left regular po-Γ-semigroups. Kangweon-Kyungki Math. Jour. **6**(2), 149–154 (1998)
14. Mordeson, J.N., Malik, D.S., Kuroki, N.: Fuzzy Semigroups. Springer, Berlin (2003)
15. Pal, P., Sardar, S.K., Majumder, S.K., Davvaz, B.: Regularity of Po-Γ-Semigroups in terms of fuzzy subsemigroups and fuzzy bi-ideals. (Communicated)
16. Rosenfeld, A.: Fuzzy groups. J. Math. Anal. Appl. **35**, 512–517 (1971)
17. Sardar, S.K., Majumder, S.K.: On fuzzy ideals in Γ-semigroups. Int. J. Algebra **3**(16), 775–784 (2009)
18. Sardar, S.K., Majumder, S.K., Mandal, D.: A note on characterization of prime ideals of Γ-semigroups in terms of fuzzy subsets. Int. J. Contemp. Math. Sci. **4**(30), 1465–1472 (2009)
19. Sardar, S.K., Majumder, S.K.: A note on characterization of semiprime ideals of Γ-semigroups in terms of fuzzy subsets. Int. J. Pure Appl. Math. **56**(3), 451–458 (2009)
20. Sardar, S.K., Pal, P., Majumder, S.K., Mukherjee, R.: On fuzzy ideals of po-Γ-semigroups. Int. Electron. J. Pure Appl. Math. **5**(2), 79–93 (2012)
21. Sardar, S.K., Pal, P.: Prime fuzzy ideals, weakly prime fuzzy ideals of Γ-semigroups. Fuzzy Inf. Eng. **5**(4), 383–397 (2013)
22. Sen, M.K.: On Γ-semigroup. Algebra and its Application, pp. 301–308. New Delhi (1981) (Lecture Notes in Pure and Applied Mathematics, vol. 91. Dekker Publication, New York (1984))
23. Sen, M.K., Saha, N.K.: On Γ-semigroup I. Bull. Calcutta Math. Soc. **78**, 180–186 (1986)
24. Sen, M.K., Seth, A.: On po-Γ-semigroups. Bull. Calcutta Math. Soc. **85**, 445–450 (1993)
25. Siripitukdet, M., Iampan, A.: On minimal and maximal ordered left ideals in po-Γ-semigroups. Thai J. Math. **2**(2), 275–282 (2004)
26. Xie, X.Y.: On prime fuzzy ideals of a semigroup. J. Fuzzy Math. **8**, 231–241 (2000)
27. Xie, X.Y., Tang, J.: On fuzzy radicals and prime fuzzy ideals of ordered semigroups. Inf. Sci. **178**(22), 4357–4374 (2008)
28. Zadeh, L.A.: Fuzzy sets. Inf. Control **8**, 338–353 (1965)

29. [illegible]

Radicals and Ideals of Affine Near-Semirings Over Brandt Semigroups

Jitender Kumar and K.V. Krishna

Abstract This work obtains all the right ideals, radicals, congruences, and ideals of the affine near-semirings over Brandt semigroups.

Keywords Near-semirings · Ideals · Radicals · Brandt semigroup

Mathematics Subject Classification (2000) Primary 16Y99 · Secondary 16N80, 20M11

1 Introduction

An algebraic structure $(\mathcal{N}, +, \cdot)$ with two binary operations $+$ and $\cdot$ is said to be a near-semiring if $(\mathcal{N}, +)$ and $(\mathcal{N}, \cdot)$ are semigroups and $\cdot$ is one-sided, say left, distributive over $+$, i.e. $a \cdot (b + c) = a \cdot b + a \cdot c$, for all $a, b, c \in \mathcal{N}$. Typical examples of near-semirings are of the form $M(S)$, the set of all mappings on a semigroup S. Near-semirings are not only the natural generalization of semirings and near-rings, but also they have very prominent applications in computer science. To name a few: process algebras by Bergstra and Klop [1], and domain axioms in near-semirings by Struth and Desharnais [3].

Near-semirings were introduced by van Hoorn and van Rootselaar as a generalization of near-rings [11]. In [10], van Hoorn generalized the concept of Jacobson radical of rings to zero-symmetric near-semirings. These radicals also generalize the radicals of near-rings by Betsch [2]. In this context, van Hoorn introduced 14 radicals of zero-symmetric near-semiring and studied some relation between them. The properties of these radicals are further investigated in the literature (e.g., [5, 12]). Krishna and Chatterjee developed a radical (which is similar to the Jacobson

J. Kumar (✉) · K.V. Krishna
Department of Mathematics, Indian Institute of Technology Guwahati, Guwahati, India
e-mail: jitender@alumni.iitg.ac.in

K.V. Krishna
e-mail: kvk@iitg.ac.in

P.G. Romeo et al. (eds.), *Semigroups, Algebras and Operator Theory*,
Springer Proceedings in Mathematics & Statistics 142,
DOI 10.1007/978-81-322-2488-4_10

radical of rings) for a special class of near-semirings to test the minimality of linear sequential machines in [6].

In this paper, we study the ideals and radicals of the zero-symmetric affine near-semiring over a Brandt semigroup. First, we present the necessary background material in Sect. 2. For the near-semiring under consideration, we obtain the right ideals in Sect. 3 and ascertain all radicals in Sect. 4. Further, we determine all its congruences and consequently obtain its ideals in Sect. 5.

2 Preliminaries

In this section, we provide a necessary background material through two subsections. One is to present the notions of near-semirings, and their ideals and radicals. In the second subsection, we recall the notion of the affine near-semiring over a Brandt semigroup. We also utilize this section to fix our notations which used throughout the work.

2.1 A Near-Semiring and Its Radicals

In this subsection, we recall some necessary fundamentals of near-semirings from [5, 10, 11].

Definition 2.1 An algebraic structure $(\mathcal{N}, +, \cdot)$ is said to be a *near-semiring* if

(1) $(\mathcal{N}, +)$ is a semigroup,
(2) $(\mathcal{N}, \cdot)$ is a semigroup, and
(3) $a \cdot (b + c) = a \cdot b + a \cdot c$, for all $a, b, c \in \mathcal{N}$.

Furthermore, if there is an element $0 \in \mathcal{N}$ such that

(4) $a + 0 = 0 + a = a$ for all $a \in \mathcal{N}$, and
(5) $a \cdot 0 = 0 \cdot a = 0$ for all $a \in \mathcal{N}$,

then $(\mathcal{N}, +, \cdot)$ is called a *zero-symmetric near-semiring*.

Example 2.2 Let $(S, +)$ be a semigroup and $M(S)$ be the set of all functions on S. The algebraic structure $(M(S), +, \circ)$ is a near-semiring, where $+$ is pointwise addition and $\circ$ is composition of mappings, i.e., for $x \in S$ and $f, g \in M(S)$,

$$x(f + g) = xf + xg \quad \text{and} \quad x(f \circ g) = (xf)g.$$

We write an argument of a function on its left, e.g., xf is the value of a function f at an argument x. We always denote the composition $f \circ g$ by fg. The notions of homomorphism and subnear-semiring of a near-semiring can be defined in a routine way.

Definition 2.3 Let $\mathcal{N}$ be a zero-symmetric near-semiring. A semigroup $(S, +)$ with identity 0_S is said to be an $\mathcal{N}$*-semigroup* if there exists a composition

$$(s, a) \mapsto sa : S \times \mathcal{N} \longrightarrow S$$

such that, for all $a, b \in \mathcal{N}$ and $s \in S$,

(1) $s(a + b) = sa + sb$,
(2) $s(ab) = (sa)b$, and
(3) $s0 = 0_S$.

Note that the semigroup $(\mathcal{N}, +)$ of a near-semiring $(\mathcal{N}, +, \cdot)$ is an $\mathcal{N}$-semigroup. We denote this $\mathcal{N}$-semigroup by $\mathcal{N}^+$.

Definition 2.4 Let S be an $\mathcal{N}$-semigroup. A semigroup congruence $\sim_r$ of S is said to be a *congruence of* $\mathcal{N}$*-semigroup* S, if for all $s, t \in S$ and $a \in \mathcal{N}$,

$$s \sim_r t \Longrightarrow sa \sim_r ta.$$

Definition 2.5 An $\mathcal{N}$*-morphism* of an $\mathcal{N}$-semigroup S is a semigroup homomorphism ϕ of S into an $\mathcal{N}$-semigroup S' such that

$$(sa)\phi = (s\phi)a$$

for all $a \in \mathcal{N}$ and $s \in S$. The kernel of an $\mathcal{N}$-morphism is called an $\mathcal{N}$*-kernel* of an $\mathcal{N}$-semigroup S. A subsemigroup T of an $\mathcal{N}$-semigroup S is said to be $\mathcal{N}$-*subsemigroup* of S if and only if $0_S \in T$ and $T\mathcal{N} \subseteq T$.

Definition 2.6 The kernel of a homomorphism of $\mathcal{N}$ is called an *ideal* of $\mathcal{N}$. The $\mathcal{N}$-kernels of the $\mathcal{N}$-semigroup $\mathcal{N}^+$ are called *right ideals* of $\mathcal{N}$.

One may refer to [10, 11] for a few other notions, viz. strong ideal, modular right ideal and λ-modular right ideal, a special congruence r''_Δ associated to a normal subsemigroup Δ of a semigroup S, and, for various (ν, μ), the $\mathcal{N}$-semigroups of type (ν, μ). The homomorphism corresponding to r''_Δ is denoted by λ_Δ.

Definition 2.7 Let s be an element of an $\mathcal{N}$-semigroup S. The annihilator of s, denoted by $A(s)$, defined by the set $\{a \in \mathcal{N} : sa = 0_S\}$. Further, for a subset T of S, the annihilator of T is

$$A(T) = \bigcap_{s \in T} A(s) = \{a \in \mathcal{N} : Ta = 0_S\}.$$

Theorem 2.8 ([5]) *The annihilator* $A(S)$ *of an* $\mathcal{N}$*-semigroup* S *is an ideal of* $\mathcal{N}$.

We now recall the notions of various radicals in the following definition.

Definition 2.9 ([10]) Let $\mathcal{N}$ be a zero-symmetric near-semiring.

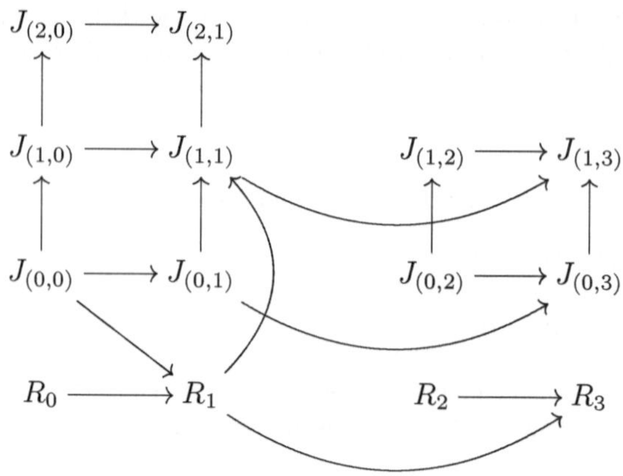

Fig. 1 Relation between various radicals of a near-semiring

(1) For $\nu = 0, 1$ with $\mu = 0, 1, 2, 3$ and $\nu = 2$ with $\mu = 0, 1$

$$J_{(\nu,\mu)}(\mathcal{N}) = \bigcap_{S \text{ is of type}(\nu,\mu)} A(S).$$

(2) $R_0(\mathcal{N})$ is the intersection of all maximal modular right ideals of $\mathcal{N}$.
(3) $R_1(\mathcal{N})$ is the intersection of all modular maximal right ideals of $\mathcal{N}$.
(4) $R_2(\mathcal{N})$ is the intersection of all maximal λ-modular right ideals of $\mathcal{N}$.
(5) $R_3(\mathcal{N})$ is the intersection of all λ-modular maximal right ideals of $\mathcal{N}$.

In any case, the empty intersection of subsets of $\mathcal{N}$ is $\mathcal{N}$. The relations between these radicals are given in Fig. 1, where $A \to B$ means $A \subset B$.

Remark 2.10 ([2, 4, 9]) If $\mathcal{N}$ is a near-ring, then $J_{(0,\mu)}(\mathcal{N})$, $\mu = 0, 1, 2, 3$ are the radical $J_0(\mathcal{N})$; $J_{(1,\mu)}(\mathcal{N})$, $\mu = 0, 1, 2, 3$ are the radical $J_1(\mathcal{N})$; $J_{(2,\mu)}(\mathcal{N})$, $\mu = 0, 1$, are the radical $J_2(\mathcal{N})$; and $R_\nu(\mathcal{N})$, $\nu = 0, 1, 2, 3$ are the radical $D(\mathcal{N})$ of Betsch. Further, if $\mathcal{N}$ is a ring, then all the 14 radicals are the radical of Jacobson.

Definition 2.11 A zero-symmetric near-semiring $\mathcal{N}$ is called (ν, μ)-*primitive* if $\mathcal{N}$ has an $\mathcal{N}$-semigroup S of type (ν, μ) with $A(S) = \{0\}$.

2.2 An Affine Near-Semiring Over a Brandt Semigroup

In this subsection, we present the necessary fundamentals of affine near-semirings over Brandt semigroups. For more details one may refer to [7, 8].

Let $(S, +)$ be a semigroup. An element $f \in M(S)$ is said to be an *affine map* if $f = g + h$, for some endomorphism g and a constant map h on S. The set of all affine mappings over S, denoted by $\mathrm{Aff}(S)$, need not be a subnear-semiring of $M(S)$.

The *affine near-semiring*, denoted by $A^+(S)$, is the subnear-semiring generated by Aff(S) in $M(S)$. Indeed, the subsemigroup of $(M(S), +)$ generated by Aff(S) equals $(A^+(S), +)$ (cf. [6, Corollary 1]). If $(S, +)$ is commutative, then Aff(S) is a subnear-semiring of $M(S)$ so that Aff$(S) = A^+(S)$.

Definition 2.12 For any integer $n \geq 1$, let $[n] = \{1, 2, \ldots, n\}$. The semigroup $(B_n, +)$, where $B_n = ([n] \times [n]) \cup \{\vartheta\}$ and the operation $+$ is given by

$$(i, j) + (k, l) = \begin{cases} (i, l) & \text{if} j = k; \\ \vartheta & \text{if} j \neq k \end{cases}$$

and, for all $\alpha \in B_n$, $\alpha + \vartheta = \vartheta + \alpha = \vartheta$, is known as *Brandt semigroup*. Note that ϑ is the (two sided) zero element in B_n.

Let ϑ be the zero element of the semigroup $(S, +)$. For $f \in M(S)$, the *support of* f, denoted by supp(f), is defined by the set

$$\text{supp}(f) = \{\alpha \in S \mid \alpha f \neq \vartheta\}.$$

A function $f \in M(S)$ is said to be of *k-support* if the cardinality of supp(f) is k, i.e. $|\text{supp}(f)| = k$. If $k = |S|$ (or $k = 1$), then f is said to be of *full support* (or *singleton support*, respectively). For $X \subseteq M(S)$, we write X_k to denote the set of all mappings of k-support in X, i.e.

$$X_k = \{f \in X \mid f \text{ is of } k\text{-support}\}.$$

For ease of reference, we continue to use the following notations for the elements of $M(B_n)$, as given in [8].

Notation 2.13

(1) For $c \in B_n$, the constant map that sends all the elements of B_n to c is denoted by ξ_c. The set of all constant maps over B_n is denoted by $\mathcal{C}_{B_n}$.
(2) For $k, l, p, q \in [n]$, the singleton support map that sends (k, l) to (p, q) is denoted by ${}^{(k,l)}\zeta_{(p,q)}$.
(3) For $p, q \in [n]$, the n-support map that sends (i, p) (where $1 \leq i \leq n$) to $(i\sigma, q)$ using a permutation $\sigma \in S_n$ is denoted by $(p, q; \sigma)$. We denote the identity permutation on $[n]$ by id.

Note that $A^+(B_1) = \{(1, 1; id)\} \cup \mathcal{C}_{B_1}$. For $n \geq 2$, the elements of $A^+(B_n)$ are given by the following theorem.

Theorem 2.14 ([8]) *For* $n \geq 2$, $A^+(B_n)$ *precisely contains* $(n! + 1)n^2 + n^4 + 1$ *elements with the following breakup.*

(1) All the $n^2 + 1$ *constant maps.*
(2) All the n^4 *singleton support maps.*

(3) *The remaining* $(n!)n^2$ *elements are the n-support maps of the form* $(p, q; \sigma)$, *where* $p, q \in [n]$ *and* $\sigma \in S_n$.

We are ready to investigate the radicals and ideals of $A^+(B_n)$—the affine near-semiring over a Brandt semigroup. Since the radicals are defined in the context of zero-symmetric near-semirings, we extend the semigroup reduct $(A^+(B_n), +)$ to monoid by adjoining 0 and make the resultant near-semiring zero-symmetric. In what follows, the zero-symmetric affine near-semiring $A^+(B_n) \cup \{0\}$ is denoted by $\mathcal{N}$, i.e.

(1) $(\mathcal{N}, +)$ is a monoid with identity element 0,
(2) $(\mathcal{N}, \circ)$ is a semigroup,
(3) $0f = f0 = 0$, for all $f \in \mathcal{N}$, and
(4) $f(g + h) = fg + fh$, for all $f, g, h \in \mathcal{N}$.

In this work, a nontrivial congruence of an algebraic structure is meant to be a congruence which is neither equality nor universal relation.

3 Right Ideals

In this section, we obtain all the right ideals of the affine near-semiring $\mathcal{N}$ by ascertaining the concerning congruences of $\mathcal{N}$-semigroups. We begin with the following lemma.

Lemma 3.1 *Let* $\sim$ *be a nontrivial congruence over the semigroup* $(\mathcal{N}, +)$ *and* $f \in A^+(B_n)_{n^2+1}$. *If* $f \sim \xi_\vartheta$, *then* $\sim = (A^+(B_n) \times A^+(B_n)) \cup \{(0, 0)\}$.

Proof First, note that $(A^+(B_n) \times A^+(B_n)) \cup \{(0, 0)\}$ is a congruence relation of the semigroup $(\mathcal{N}, +)$. Let $f = \xi_{(p_0,q_0)}$ and $\xi_{(p,q)}$ be an arbitrary full support map. Since

$$\xi_{(p,q)} = \xi_{(p,p_0)} + \xi_{(p_0,q_0)} + \xi_{(q_0,q)} \sim \xi_{(p,p_0)} + \xi_\vartheta + \xi_{(q_0,q)} = \xi_\vartheta,$$

we have $\xi_{(p,q)} \sim \xi_\vartheta$ for all $p, q \in [n]$. Further, given an arbitrary n-support map $(k, l; \sigma)$, since $\xi_{(p,l)} \sim \xi_\vartheta$, we have

$$(k, l; \sigma) = (k, p; \sigma) + \xi_{(p,l)} \sim (k, p; \sigma) + \xi_\vartheta = \xi_\vartheta.$$

Thus, all n-support maps are related to ξ_ϑ under $\sim$. Similarly, given an arbitrary ${}^{(k,l)}\zeta_{(p,q)} \in A^+(B_n)_1$, since $\xi_{(p,q)} \sim \xi_\vartheta$, for $\sigma \in S_n$ such that $k\sigma = q$, we have

$${}^{(k,l)}\zeta_{(p,q)} = \xi_{(p,q)} + (l, q; \sigma) \sim \xi_\vartheta + (l, q; \sigma) = \xi_\vartheta.$$

Hence, all elements of $A^+(B_n)$ are related to each other under $\sim$. □

Now, using Lemma 3.1, we determine the right ideals of $\mathcal{N}$ in the following theorem.

Theorem 3.2 *$\mathcal{N}$ and $\{0\}$ are only the right ideals of $\mathcal{N}$.*

Proof Let $I \neq \{0\}$ be a right ideal of $\mathcal{N}$ so that $I = \ker\varphi$, where $\varphi : \mathcal{N}^+ \longrightarrow S$ is an $\mathcal{N}$-morphism. Note that $I = [0]_{\sim_r}$, where $\sim_r$ is the congruence over the $\mathcal{N}$-semigroup $\mathcal{N}^+$ defined by $a \sim_r b$ if and only if $a\varphi = b\varphi$, i.e. the relation $\sim_r$ on $\mathcal{N}$ is compatible with respect to $+$ and if $a \sim_r b$ then $ac \sim_r bc$ for all $c \in \mathcal{N}$.

Let f be a nonzero element of $\mathcal{N}$ such that $f \sim_r 0$. First, note that

$$\xi_\vartheta = f\xi_\vartheta \sim_r 0\xi_\vartheta = 0.$$

Further, for any full support map $\xi_{(p,q)}$, we have

$$\xi_{(p,q)} = f\xi_{(p,q)} \sim_r 0\xi_{(p,q)} = 0$$

so that, by transitivity, $\xi_{(p,q)} \sim_r \xi_\vartheta$. Hence, by Lemma 3.1, $\sim_r = \mathcal{N} \times \mathcal{N}$ so that $I = \mathcal{N}$. □

Remark 3.3 The ideal $\{0\}$ is the maximal right ideal of $\mathcal{N}$.

4 Radicals

In order to obtain the radicals of the affine near-semiring $\mathcal{N}$, in this section, we first identify an $\mathcal{N}$-semigroup which satisfies the criteria of all types of $\mathcal{N}$-semigroups by van Hoorn. Using the $\mathcal{N}$-semigroup, we ascertain the radicals of $\mathcal{N}$. Further, we observe that the near-semiring $\mathcal{N}$ is (ν, μ)-primitive (cf. Theorem 4.3).

Consider the subsemigroup $\mathcal{C} = \mathcal{C}_{B_n} \cup \{0\}$ of $(\mathcal{N}, +)$ and observe that $\mathcal{C}$ is an $\mathcal{N}$-semigroup with respect to the multiplication in $\mathcal{N}$. The following properties of the $\mathcal{N}$-semigroup $\mathcal{C}$ are useful.

Lemma 4.1

(1) *Every nonzero element of $\mathcal{C}$ is a generator. Moreover, the $\mathcal{N}$-semigroup $\mathcal{C}$ is strongly monogenic and $A(g) = \{0\}$ for all $g \in \mathcal{C}\setminus\{0\}$.*
(2) *The subsemigroup $\{0\}$ is the maximal $\mathcal{N}$-subsemigroup of $\mathcal{C}$.*
(3) *The $\mathcal{N}$-semigroup $\mathcal{C}$ is irreducible.*

Proof

(1) Let $g \in \mathcal{C}_{B_n}$. Note that $g\mathcal{N} \subseteq \mathcal{C}$ because the product of a constant map with any map is a constant map. Conversely, for $f \in \mathcal{C}$, since $gf = f$, we have $g\mathcal{N} = \mathcal{C}$ for all $g \in \mathcal{C}\setminus\{0\}$. Further, since $0\mathcal{N} = \{0\}$ and $\mathcal{C}\mathcal{N} = \mathcal{C} \neq \{0\}$. Hence, $\mathcal{C}$ is strongly monogenic.

(2) We show that the semigroups $\mathcal{C}$ and $\{0\}$ are the only $\mathcal{N}$-subsemigroups of $\mathcal{C}$. Let T be an $\mathcal{N}$-subsemigroup of $\mathcal{C}$ such that $\{0\} \neq T \subsetneq \mathcal{C}$. Then there exist $f(\neq 0) \in T$ and $g \in \mathcal{C} \setminus T$. Since $fg = g \notin T$, we have $T\mathcal{N} \nsubseteq T$; a contradiction to T is an $\mathcal{N}$-subsemigroup. Hence, the result.

(3) By Lemma 4.1(1), the $\mathcal{N}$-semigroup $\mathcal{C}$ is monogenic with any nonzero element g as generator such that $A(g) = \{0\}$; thus, $A(g)$ is maximal right ideal in $\mathcal{N}$ (cf. Remark 3.3). Hence, by [10, Theorem 8], $\mathcal{C}$ is irreducible. □

Remark 4.2 Since a strongly monogenic $\mathcal{N}$-semigroup is monogenic we have, for $\mu = 0, 1, 2, 3$, an $\mathcal{N}$-semigroup of type $(1, \mu)$ is of type $(0, \mu)$.

Theorem 4.3 *For $\nu = 0, 1$ with $\mu = 0, 1, 2, 3$ and $\nu = 2$ with $\mu = 0, 1$, we have the following.*

(1) The $\mathcal{N}$-semigroup $\mathcal{C}$ is of type (ν, μ) with $A(\mathcal{C}) = 0$.
(2) The near-semiring $\mathcal{N}$ is (ν, μ)-primitive for all ν and μ.
(3) $J_{(\nu,\mu)}(\mathcal{N}) = \{0\}$ for all ν and μ.

Proof In view of Remark 4.2, we prove (1) in the following cases.

Type $(1, \mu)$ Note that, by Lemma 4.1(1), the $\mathcal{N}$-semigroup $\mathcal{C}$ is strongly monogenic.

(i) By Lemma 4.1(3), we have $\mathcal{C}$ is irreducible. Hence, $\mathcal{C}$ is of type $(1, 0)$.
(ii) By Lemma 4.1(1) and Remark 3.3, for any generator g, $A(g)$ is a maximal right ideal. Hence, $\mathcal{C}$ is of type $(1, 1)$.
(iii) Note that the ideal $\{0\}$ is strong right ideal so that for any generator g, $A(g)$ is a strong maximal right ideal (see ii above). Further, note that $A(g)$ is a maximal strong right ideal (cf. Remark 3.3). Hence, $\mathcal{C}$ is of type $(1, 2)$ and $(1, 3)$.

Type $(2, \mu)$ Since $\mathcal{C}$ is monogenic and, for any generator g of $\mathcal{C}$, $A(g)$ is a maximal $\mathcal{N}$-subsemigroup of $\mathcal{C}$ (cf. Lemma 4.1(1) and Lemma 4.1(2)). Thus, $\mathcal{C}$ is of type $(2, 1)$. By [10, Theorem 9], every $\mathcal{N}$-semigroup of type $(2, 1)$ will be of type $(2, 0)$. Hence, $\mathcal{C}$ is of type $(2, 0)$.

Proofs for (2) and (3) follow from (1). □

Theorem 4.4 *For $\nu = 0, 1$, we have $R_\nu(\mathcal{N}) = \{0\}$.*

Proof In view of Fig. 1, we prove the result by showing that the right ideal $\{0\}$ is a modular maximal right ideal. By Lemma 4.1(1), the $\mathcal{N}$-semigroup $\mathcal{C}$ is monogenic and has a generator g such that $A(g) = \{0\}$. Hence, the right ideal $\{0\}$ is modular (cf. [10, Theorem 7]). Further, since $\{0\}$ is a maximal right ideal (cf. Remark 3.3), we have $\{0\}$ is a modular maximal right ideal. □

Theorem 4.5 *For $\nu = 2, 3$, we have $R_\nu(\mathcal{N}) = \mathcal{N}$.*

Proof In view of Fig. 1 and Theorem 3.2, we prove that the homomorphism $\lambda_{\{0\}}$ is not modular. Note that the congruence relation $r''_{\{0\}}$ is the equality relation on $(\mathcal{N}, +)$, where $r''_{\{0\}}$ is the transitive closure of the two-sided stable reflexive and symmetric relation relation $r_{\{0\}}$ associated with a normal subsemigroup $\{0\}$ of the semigroup $(\mathcal{N}, +)$. Consequently, the semigroup homomorphism $\lambda_{\{0\}}$ is an identity map on $(\mathcal{N}, +)$. If the morphism $\lambda_{\{0\}}$ is modular, then there is an element $u \in \mathcal{N}$ such that $x = ux$ for all $x \in \mathcal{N}$, but there is no left identity element in $\mathcal{N}$. Consequently, $\lambda_{\{0\}}$ is not modular. Thus, there is no maximal λ-modular right ideal. Hence, for $\nu = 2, 3$, we have $R_\nu(\mathcal{N}) = \mathcal{N}$. □

5 Ideals

In this section, we prove that there is only one nontrivial congruence relation on $\mathcal{N}$ (cf. Theorem 5.1). Consequently, all the ideals of $\mathcal{N}$ are determined.

Theorem 5.1 *The near-semiring $\mathcal{N}$ has precisely the following congruences.*

(1) Equality relation
(2) $\mathcal{N} \times \mathcal{N}$
(3) $(A^+(B_n) \times A^+(B_n)) \cup \{(0, 0)\}$

Hence, $\mathcal{N}$ and $\{0\}$ are the only ideals of the near-semiring $\mathcal{N}$.

Proof In the sequel, we prove the theorem through the following claims.

Claim 1 Let $\sim$ be a nontrivial congruence over the near-semiring $\mathcal{N}$ and $f \in \mathcal{N} \setminus \{0, \xi_\vartheta\}$. If $f \sim \xi_\vartheta$, then $\sim = (A^+(B_n) \times A^+(B_n)) \cup \{(0, 0)\}$.

Proof First, note that $(A^+(B_n) \times A^+(B_n)) \cup \{(0, 0)\}$ is a congruence relation of the near-semiring $\mathcal{N}$. If $f \in A^+(B_n)_{n^2+1}$, since $\sim$ is a congruence of the semigroup $(\mathcal{N}, +)$, by Lemma 3.1, we have the result. Otherwise, we reduce the problem to Lemma 3.1 in the following cases.

Case 1.1 f is of singleton support. Let $f = {}^{(k,l)}\zeta_{(p,q)}$. Since ${}^{(k,l)}\zeta_{(p,q)} \sim \xi_\vartheta$ we have

$$\xi_{(k,l)} \, {}^{(k,l)}\zeta_{(p,q)} \sim \xi_{(k,l)}\xi_\vartheta$$

so that $\xi_{(p,q)} \sim \xi_\vartheta$.

Case 1.2 f is of n-support. Let $f = (p, q; \sigma)$. Since $(p, q; \sigma) \sim \xi_\vartheta$ we have

$$\xi_{(k,p)}(p, q; \sigma) \sim \xi_{(k,p)}\xi_\vartheta$$

so that $\xi_{(k\sigma,q)} \sim \xi_\vartheta$.

Claim 2 If two nonzero elements are in one class under a nontrivial congruence over $\mathcal{N}$, then the congruence is $(A^+(B_n) \times A^+(B_n)) \cup \{(0, 0)\}$.

Proof Let $f, g \in \mathcal{N} \backslash \{0\}$ such that $f \sim g$ under a congruence $\sim$ over $\mathcal{N}$. If f or g is equal to ξ_ϑ, then by *Claim 1*, we have the result. Otherwise, we consider the following six cases classified by the supports of f and g. In each case, we show that there is an element $h \in A^+(B_n) \backslash \{\xi_\vartheta\}$ such that $h \sim \xi_\vartheta$ so that the result follows from *Claim 1*.

Case 2.1 $f, g \in A^+(B_n)_1$. Let $f = {}^{(i,j)}\zeta_{(k,l)}$ and $g = {}^{(s,t)}\zeta_{(u,v)}$. If $(i, j) \neq (s, t)$, we have

$$\xi_\vartheta = {}^{(i,j)}\zeta_{(k,l)} + {}^{(s,t)}\zeta_{(v,v)} \sim {}^{(s,t)}\zeta_{(u,v)} + {}^{(s,t)}\zeta_{(v,v)} = {}^{(s,t)}\zeta_{(u,v)}.$$

Otherwise, $(i, j) = (s, t)$ so that $(k, l) \neq (u, v)$. Now, if $k \neq u$, then we have

$${}^{(i,j)}\zeta_{(k,l)} = {}^{(i,j)}\zeta_{(k,k)} + {}^{(i,j)}\zeta_{(k,l)} \sim {}^{(i,j)}\zeta_{(k,k)} + {}^{(i,j)}\zeta_{(u,v)} = \xi_\vartheta.$$

Similarly, if $l \neq v$, we have

$$\xi_\vartheta = {}^{(i,j)}\zeta_{(k,l)} + {}^{(i,j)}\zeta_{(v,v)} \sim {}^{(i,j)}\zeta_{(u,v)} + {}^{(i,j)}\zeta_{(v,v)} = {}^{(i,j)}\zeta_{(u,v)}.$$

Case 2.2 $f, g \in A^+(B_n)_{n^2+1}$. Let $f = \xi_{(k,l)}$ and $g = \xi_{(u,v)}$. By considering full support maps whose images are the same as in various subcases of *Case 1*, we can show that there is an element in $A^+(B_n) \backslash \{\xi_\vartheta\}$ that is related to ξ_ϑ under $\sim$.

Case 2.3 $f, g \in A^+(B_n)_n$. Let $f = (i, j; \sigma)$ and $g = (k, l; \rho)$. If $l \neq j$, then

$$(i, j; \sigma) = (i, j; \sigma) + \xi_{(j,j)} \sim (k, l; \rho) + \xi_{(j,j)} = \xi_\vartheta.$$

Otherwise, we have $(i, j; \sigma) \sim (k, j; \rho)$. Now, if $i \neq k$, then

$$\xi_\vartheta = (k, k; id)(i, j; \sigma) \sim (k, k; id)(k, j; \rho) = (k, j; \rho).$$

In case $i = k$, we have $\sigma \neq \rho$. Thus, there exists $t \in [n]$ such that $t\sigma \neq t\rho$. Now, $(i, j; \sigma) \sim (i, j; \rho)$ implies $\xi_{(k,i)}(i, j; \sigma) \sim \xi_{(k,i)}(i, j; \rho)$, i.e. $\xi_{(k\sigma,j)} \sim \xi_{(k\rho,j)}$. Consequently,

$$\xi_{(k\sigma,j)} = \xi_{(k\sigma,k\sigma)} + \xi_{(k\sigma,j)} \sim \xi_{(k\sigma,k\sigma)} + \xi_{(k\rho,j)} = \xi_\vartheta.$$

Case 2.4 $f \in A^+(B_n)_1, g \in A^+(B_n)_{n^2+1}$. Let $f = {}^{(k,l)}\zeta_{(p,q)}$ and $g = \xi_{(i,j)}$. Now, for $(s, t) \neq (k, l)$, we have

$$\xi_\vartheta = \xi_{(s,t)} f \sim \xi_{(s,t)} g = \xi_{(i,j)}.$$

Case 2.5 $f \in A^+(B_n)_{n^2+1}, g \in A^+(B_n)_n$. Let $f = \xi_{(p,q)}$ and $g = (i, j; \sigma)$. Now, for $l \neq i$, we have

$$^{(k,l)}\zeta_{(p,q)} = {}^{(k,l)}\zeta_{(p,p)} + f \sim {}^{(k,l)}\zeta_{(p,p)} + g = \xi_\vartheta.$$

Case 2.6 $f \in A^+(B_n)_1, g \in A^+(B_n)_n$. Let $f = {}^{(k,l)}\zeta_{(p,q)}$ and $g = (i, j; \sigma)$. Now, for $l \neq i$, we have

$$\xi_\vartheta = \xi_{(i,i)} f \sim \xi_{(i,i)} g = \xi_{(i\sigma,j)}.$$ □

References

1. Bergstra, J.A., Klop, J.W.: An Introduction to Process Algebra. Cambridge Tracts in Theoretically Computer Science, vol. 17. Cambridge University Press, Cambridge (1990)
2. Betsch, G.: Struktursätze für Fastringe. Inaugural-Dissertation. Eberhard-Karls-Universität zu Tübingen (1963)
3. Desharnais, J., Struth, G.: Domain axioms for a family of near-semirings. In: AMAST, pp. 330–345 (2008)
4. Jacobson, N.: Structure of rings. Am. Math. Soc. Colloq. Publ., **37** (1964). (Revised edition. American Mathematical Society, Providence, R.I.)
5. Krishna, K.V.: Near-Semirings: Theory and Application. Ph.D. thesis, IIT Delhi, New Delhi (2005)
6. Krishna, K.V., Chatterjee, N.: A necessary condition to test the minimality of generalized linear sequential machines using the theory of near-semirings. Algebra Discret. Math. **3**, 30–45 (2005)
7. Kumar, J.: Affine Near-Semirings over Brandt Semigroups. Ph.D. thesis, IIT Guwahati (2014)
8. Kumar, J., Krishna, K.V.: Affine near-semirings over Brandt semigroups. Commun. Algebra **42**(12), 5152–5169 (2014)
9. Pilz, G.: Near-Rings: The Theory and its Applications. North-Holland Mathematics Studies, vol. 23. North-Holland Publishing Company (1983)
10. van Hoorn, W.G.: Some generalisations of the Jacobson radical for semi-near rings and semi-rings. Math. Z. **118**, 69–82 (1970)
11. van Hoorn, W.G., van Rootselaar, B.: Fundamental notions in the theory of seminearrings. Compos. Math. **18**, 65–78 (1967)
12. Zulfiqar, M.: A note on radicals of seminear-rings. Novi Sad J. Math. **39**(1), 65–68 (2009)

Operator Approximation

Balmohan V. Limaye

Abstract We present an introduction to operator approximation theory. Let T be a bounded linear operator on a Banach space X over $\mathbb{C}$. In order to find approximate solutions of (i) the operator equation $zx - Tx = y$, where $z \in \mathbb{C}$ and $y \in X$ are given, and (ii) the eigenvalue problem $T\varphi = \lambda\varphi$, where $\lambda \in \mathbb{C}$ and $0 \neq \varphi \in X$, one approximates the operator T by a sequence (T_n) of bounded linear operators on X. We consider pointwise convergence, norm convergence, and nu convergence of (T_n) to T. We give several examples to illustrate possible scenarios. In most classical methods of approximation, each T_n is of finite rank. We give a canonical procedure for reducing problems involving finite rank operators to problems involving matrix computations.

Keywords Operator equation · Eigenvalue problem · Resolvent set · Resolvent operator · Spectrum · Spectral projection · Approximate solution · Error estimate · Iterated version · Pointwise convergence · Norm convergence · Nu convergence · Integral operator · Degenerate kernel approximation · Projection approximation · Sloan approximation · Galerkin approximation · Nyström approximation · Interpolatory projection · Finite rank operator

Mathematics Subject Classification (2010) 47A58 · 47-02 · 47A75 · 65J10

1 Introduction

Let X be a Banach Space over $\mathbb{C}$, and let $T : X \to X$ be a bounded linear operator on X, that is, $T \in BL(X)$. We address the following problem:

"Given $y \in X$ and $z \in \mathbb{C}$, find $x \in X$ such that $zx - Tx = y$."

Case I: Let $z \in \mathbb{C}$, and suppose that for every $y \in X$ there is unique $x \in X$ such that $zx - Tx = y$. This means z is in the resolvent set $\rho(T)$ of the operator T. We

B.V. Limaye (✉)
Department of Mathematics, Indian Institute of Technology Bombay,
Powai, Mumbai 400076, India
e-mail: bvl@math.iitb.ac.in

P.G. Romeo et al. (eds.), *Semigroups, Algebras and Operator Theory*,
Springer Proceedings in Mathematics & Statistics 142,
DOI 10.1007/978-81-322-2488-4_11

let $R(z) := (zI - T)^{-1}$, the **resolvent operator** of T at z. Clearly it is linear. By the bounded inverse theorem, it is also bounded. Thus $R(z) \in BL(X)$.

Case II: Let $z \in \mathbb{C}$, and suppose that either there is $y \in X$ such that $zx - Tx \neq y$ for any $x \in X$, or there are $x_1 \neq x_2 \in X$ such that $zx_1 - Tx_1 = y = zx_2 - Tx_2$. This means z is in the **spectrum** $\sigma(T)$ of T.

We shall see later that the complex number $z = 0$ has a peculiar significance in both cases. Both problems, especially the second, can be difficult to solve when the Banach space X is infinite dimensional. As a result, one often replaces the operator T by a "simpler" operator $\widetilde{T}$ which is "nearb" T. This process gives rise to **Operator Approximation Theory** which is a well-developed branch of the so-called **Numerical Functional Analysis.** This article is an introduction to operator approximation. It is not a comprehensive survey. It will include some classical methods as well as some recent developments. Interested readers can pursue further study by consulting the references given at the end of this article.

In Sect. 2, we treat **Case I** and in Sect. 3, we treat **Case II**. In Sect. 4, we consider finite rank operators. They are often used to approximate compact operators.

We consider a sequence (T_n) in $BL(X)$ which approximates T is a sense to be made precise subsequently.

2 Solution of Operator Equations

Let $z \in \rho(T)$, that is, the operator $zI - T$ is bijective. Fix $n \in \mathbb{N}$ and suppose $z \in \rho(T_n)$, that is, the operator $zI - T_n$ is also bijective. Let $R_n(z) := (zI - T_n)^{-1}$, the resolvent operator of T_n at z.

Consider $y \in X$, and let $x \in X$ be such that $zx - Tx = y$. Consider $y_n \in X$, and let $x_n \in X$ be such that $zx_n - T_nx_n = y_n$. Then

$$\begin{aligned}(zI - T_n)(x_n - x) &= zx_n - T_nx_n + T_nx - Tx - zx + Tx \\ &= T_nx - Tx + y_n - y,\end{aligned}$$

that is,

$$x_n - x = R_n(z)(T_nx - Tx + y_n - y)$$

For $n \in \mathbb{N}$, we may choose $y_n \in X$ such that $y_n \to y$. If $T_nx \to Tx$, and if the sequence $(\|R_n(z)\|)$ is bounded, then it follows that $x_n \to x$. We may then say that (x_n) is a sequence of **approximate solutions** of the operator equation $zx - Tx = y$.

Thus we are led to consider the following mode of convergence:

Pointwise Convergence: Let $T_n(x) \to T(x)$ for every $x \in X$. We denote this by $T_n \xrightarrow{p} T$.

Next, we ask the following question: If $T_n \xrightarrow{p} T$ and $z \in \rho(T)$, will $z \in \rho(T_n)$ for all large n, and then will the sequence $(\|R_n(z)\|)$ be bounded? The answers are in the negative.

Example 2.1 Let $X := \ell^2$, the Banach space of all square-summable complex sequences.

(i) For $x := (x_1, x_2, \ldots) \in X$, let

$$Tx := (x_1, 0, 0, \ldots) \quad \text{and} \quad T_n x := (x_1, 0, \ldots, 0, -x_n, 0, 0, \ldots).$$

Then $T_n \xrightarrow{p} T$, $-1 \in \rho(T)$, but $-1 \notin \rho(T_n)$ for any $n \in \mathbb{N}$.

(ii) For $x := (x_1, x_2, \ldots) \in X$, let

$$Tx := (0, 0, \ldots) \quad \text{and} \quad T_n x := \frac{n-1}{n}(0, \ldots, 0, x_n, x_{n+1}, \ldots).$$

Then $T_n \xrightarrow{p} T$ and $1 \in \rho(T) \cap \rho(T_n)$ for every $n \in \mathbb{N}$. However, $(I - T_n)^{-1}y = (y_1, \ldots, y_{n-1}, ny_n, ny_{n+1}, \ldots)$ for $y \in X$, and so $\|R_n(1)\| = n$ for every $n \in \mathbb{N}$. Hence the sequence $(\|R_n(1)\|)$ is not bounded.

Thus we are led to consider a stronger mode of convergence:

Norm Convergence: Let $\|T_n - T\| \to 0$. We denote this by $T_n \xrightarrow{n} T$.

For $z \in \rho(T)$ and $n \in \mathbb{N}$, define $S_n(z) := (T_n - T)R(z)$.

Proposition 2.2 *Let $T_n \xrightarrow{n} T$ and $z \in \rho(T)$. Then $S_n(z) \xrightarrow{n} 0$. If $\|S_n(z)\| < 1$, then $z \in \rho(T_n)$ and*

$$\|R_n(z)\| \leq \frac{\|R(z)\|}{1 - \|S_n(z)\|}.$$

In particular, $(\|R_n(z)\|)$ is a bounded sequence.

Proof Let $z \in \rho(T)$. Then $\|S_n(z)\| \leq \|T_n - T\| \, \|R(z)\|$, and so $S_n(z) \xrightarrow{n} 0$. Also,

$$zI - T_n = \big(I - (T_n - T)(zI - T)^{-1}\big)(zI - T) = [I - S_n(z)](zI - T).$$

If $\|S_n(z)\| < 1$, then $I - S_n(z)$ is invertible and the norm of its inverse is less than or equal to $1/[1 - \|S_n(z)\|]$, and so $z \in \rho(T_n)$ and $\|R_n(z)\| \leq \|R(z)\|/[1 - \|S_n(z)\|]$. It follows that $(\|R_n(z)\|)$ is a bounded sequence. □

Corollary 2.3 *Let $T_n \xrightarrow{n} T$ and $z \in \rho(T)$. Then $z \in \rho(T_n)$ for all large n. If $zx - Tx = y$ and $zx_n - T_n x_n = y_n$, then for all large n,*

$$\|x_n - x\| \leq 2\|R(z)\|(\|T_n x - Tx\| + \|y_n - y\|).$$

Proof There is $n_0 \in \mathbb{N}$ such that $\|S_n(z)\| \leq 1/2$ for all $n \geq n_0$. Since $x_n - x = R_n(z)(T_n x - Tx + y_n - y)$ whenever $z \in \rho(T) \cap \rho(T_n)$, the result follows from Proposition 2.2. □

This result gives an error estimate for the approximate solution x_n of the operator equation $zx - Tx = y$.

Suppose $z \neq 0$. Now $x = (y + Tx)/z$. An iterated version of the approximate solution $\widetilde{x}_n$ of x is, therefore, defined by

$$\widetilde{x}_n := \frac{1}{z}(y + Tx_n).$$

Then $\widetilde{x}_n - x = T(x_n - x)/z$, and $\widetilde{x}_n$ would be a better approximation of x than x_n if $\|T(x_n - x)\| \leq \epsilon_n \|x_n - x\|$, where $\epsilon_n \to 0$. (See Remarks 3.10 and 3.12, and Theorem 3.24 of [11]).

We thus note that existence and convergence of approximate solutions to the exact solution of an operator equation is guaranteed under the norm convergence of the sequence (T_n) to T, but not under the pointwise convergence of (T_n) to T. However, it turns out that many classical sequences of operators used for approximating an operator T do not converge in the norm. We give some examples of norm convergence, and also some examples where norm convergence fails.

Example 2.4 Let $X = C([a, b])$ and for $x \in X$, consider

$$(Tx)(s) := \int_a^b k(s, t)x(t)dt, \quad s \in [a, b],$$

where the kernel $k(\cdot, \cdot)$ is continuous on $[a, b] \times [a, b]$. Then T is a compact operator.

(i) For $n \in \mathbb{N}$, let

$$k_n(s, t) := \sum_{j=1}^{n} x_{n,j}(s)\ y_{n,j}(t), \quad s, t \in [a, b],$$

be a degenerate kernel, where $x_{n,j}$ and $y_{n,j}$ are complex-valued continuous functions on $[a, b]$. Assume that $\|k_n(\cdot, \cdot) - k(\cdot, \cdot)\|_\infty \to 0$. For $x \in X$, define

$$(T_n^D x)(s) := \int_a^b k_n(s, t)x(t)dt, \quad s \in [a, b].$$

Then $\|T_n^D - T\| \leq (b-a)\|k_n(\cdot, \cdot) - k(\cdot, \cdot)\|_\infty \to 0$. This is known as the degenerate kernel approximation.

(ii) For $n \in \mathbb{N}$, let $e_{n,1}, \ldots, e_{n,n}$ be the hat functions with nodes at $t_{n,1}, \ldots, t_{n,n}$, where $t_{n,j} = a + (b - a)j/n$. Consider the interpolatory projection $\pi_n : X \to X$ given by

$$\pi_n x := \sum_{j=1}^{n} x(t_{n,j})e_{n,j}, \quad x \in X.$$

For $x \in X$, let $T_n^P x := \pi_n T x$. Then $\pi_n x \to x$ for every $x \in X$ and

$$\|T_n^P - T\| = \sup\{\|\pi_n T x - T x\| : x \in X \text{ and } \|x\| \le 1\} \to 0 \text{ as } n \to \infty,$$

since T is compact. However, if $T_n^S := T\pi_n$ and $T_n^G := \pi_n T \pi_n$, then $\|T_n^S - T\| \not\to 0$ and $\|T_n^G - T\| \not\to 0$, unless $T = 0$. (See Proposition 4.6 of [2], and the preceding comment.) On the other hand, if $T_n^K := T_n^P + T_n^S - T_n^G$, then $\|T - T_n^K\| = \|(I - \pi_n)T(I - \pi_n)\| \le \|(I - \pi_n)T\| \, \|(I - \pi_n)\| \to 0$, since T is compact and the sequence $(\|\pi_n\|)$ is bounded. The approximations T_n^P, T_n^S, and T_n^G are known as the projection approximation, the Sloan approximation, and the Galerkin approximation of T, respectively. The approximation T_n^K was introduced by Kulkarni [9]. It may be called a modified projection approximation of T.

(iii) For $n \in \mathbb{N}$, let

$$Q_n x := \sum_{j=1}^{n} w_{n,j}\, x(t_{n,j}), \quad x \in X,$$

be a quadrature formula, where $w_{n,1}, \ldots, w_{n,n}$ are the weights. Assume that $Q_n x \to \int_a^b x(t)dt$ for every $x \in X$. For $x \in X$, define

$$(T_n^N x)(s) := \sum_{j=1}^{n} w_{n,j} k(s, t_{n,j}) x(t_{n,j}), \quad s \in [a, b].$$

The approximation T_n^N is known as the the Nyström approximation of T. Then $\|T_n^N - T\| \not\to 0$ unless $T = 0$. (See Proposition 4.6 of [2], and the preceding comment).

To encompass all the above important examples, we seek a mode of convergence which is weaker than the norm convergence, but which guarantees the boundedness of the sequence $(\|R_n(z)\|)$ as in Proposition 2.2. Anselone and Moore [4], proposed "collectively compact" convergence in this connection. See also Atkinson [5–7] and Anselone [3]. The implications of this concept are captured in the following mode of convergence.

Nu Convergence: Let $T_n \xrightarrow{p} T$, $\|(T_n - T)T\| \to 0$ and $\|(T_n - T)T_n\| \to 0$. We denote this by $T_n \xrightarrow{\nu} T$.

It is based on the work of Ahues [1], Bouldin [8], and Nair [10]. In the book of Ahues et al. [2], nu convergence was defined by the following conditions: $(\|T_n\|)$ is bounded, $\|(T_n - T)T\| \to 0$ and $\|(T_n - T)T_n\| \to 0$, and was extensively used in the context of spectral approximation. Note that if $T_n \xrightarrow{p} T$, then $(\|T_n\|)$ is bounded by the uniform boundedness principle. Thus the definition of nu convergence stated above is slightly stronger.

Clearly, if $T_n \xrightarrow{n} T$, then $T_n \xrightarrow{\nu} T$, and if $T_n \xrightarrow{\nu} T$, then $T_n \xrightarrow{p} T$. More importantly, it can be shown that $T_n^S \xrightarrow{\nu} T$, $T_n^G \xrightarrow{\nu} T$ and $T_n^N \xrightarrow{\nu} T$. (See [2], parts (a) and (b) of Theorem 4.1, and part (a) of Theorem 4.5).

Proposition 2.5 *Let $T \xrightarrow{\nu} T$ and $z \in \rho(T)$ with $z \neq 0$. Then $[S_n(z)]^2 \xrightarrow{n} 0$. If $\|[S_n(z)]^2\| < 1$, then $z \in \rho(T_n)$ and*

$$\|R_n(z)\| \leq \frac{\|R(z)\|(1 + \|S_n(z)\|)}{1 - \|[S_n(z)]^2\|}.$$

In particular, $(\|R_n(z)\|)$ is bounded.

Proof Let $z \in \rho(T)$. Then

$$\begin{aligned}
[S_n(z)]^2 &= (T_n - T)R(z)(T_n - T)R(z) \\
&= (T_n - T)\Big(\frac{I + TR(z)}{z}\Big)(T_n - T)R(z) \\
&= \frac{1}{z}\big((T_n - T)^2 + (T_n - T)TR(z)(T_n - T)\big)R(z).
\end{aligned}$$

Now $\|(T_n - T)^2\| \leq \|(T_n - T)T_n\| + \|(T_n - T)T\| \to 0$ and $(\|T_n\|)$ is bounded. Hence $\|[S_n(z)]^2\| \to 0$. If $\|[S_n(z)]^2\| < 1$, then $I - S_n(z)$ is invertible and the norm of its inverse is less than or equal to $(1 + \|S_n(z)\|)/\big(1 - \|[S_n(z)]^2\|\big)$, and since $zI - T_n = (I - S_n(z))R(z)$ as in Proposition 2.2, $z \in \rho(T_n)$ and $\|R_n(z)\| \leq \|R(z)\|(1 + \|S_n(z)\|)/\big(1 - \|[S_n(z)]^2\|\big)$. It follows that $(\|R_n(z)\|)$ is a bounded sequence. □

Corollary 2.6 *Let $T_n \xrightarrow{\nu} T$ and $z \in \rho(T)$ with $z \neq 0$. Then $z \in \rho(T_n)$ for all large n. If $zx - Tx = y$ and $zx_n - T_nx_n = y_n$, then for all large n,*

$$\|x_n - x\| \leq 2\|R(z)\|\big(1 + \|R(z)\|(\alpha + \|T\|)\big)(\|T_nx - Tx\| + \|y - y_n\|),$$

where $\alpha \in \mathbb{R}$ is such that $\|T_n\| \leq \alpha$ for all $n \in \mathbb{N}$.

Proof There is $n_0 \in \mathbb{N}$ such that $\|[S_n(z)]^2\| \leq 1/2$ for all $n \geq n_0$. Also, $\|S_n(z)\| \leq (\|T_n\| + \|T\|)\|R(z)\|$ for all $n \in \mathbb{N}$. Since

$$x_n - x = R_n(z)(T_nx - Tx + y_n - y)$$

whenever $z \in \rho(T) \cap \rho(T_n)$, the result follows from Proposition 2.5. □

Thus, if $T_n \xrightarrow{\nu} T$, then x_n can be considered an approximate solution of the operator equation $zx - Tx = y$, and error estimates can be given as in the case of the norm convergence.

Before we conclude this section, we give an upper bound for $\|R_n(z)\|$ when $z \neq 0$, $z \in \rho(T) \cap \rho(T_n)$ and $\|R(z)\|\,\|(T_n - T)T_n\| < |z|$ a la Theorem 4.1.1 of Atkinson's book [6]: Since $I + R(z)T_n = \big(zI - R(z)(T_n - T)T_n\big)R_n(z)$, we obtain

$$\|R_n(z)\| \le \frac{1 + \|R(z)\| \, \|T_n\|}{|z| - \|R(z)\| \, \|(T_n - T)T_n\|}.$$

If $\|(T_n - T)T_n\| \to 0$, then there is $n_0 \in \mathbb{N}$ such that $\|R(z)\| \, \|(T_n - T)T_n\| \le |z|/2$ for all $n \ge n_0$, and so $\|R_n(z)\| \le 2(1 + \|R(z)\| \, \|T_n\|)/|z|$. Hence $(\|R_n(z)\|)$ is a bounded sequence, provided $(\|T_n\|)$ is a bounded sequence.

3 Spectral Values and Eigenvalues

The spectrum of $T \in BL(X)$ is defined by

$$\sigma(T) := \{\lambda \in \mathbb{C} : \lambda I - T \text{ is not bijective}\}.$$

Let (T_n) be a sequence in $BL(X)$ such that $T_n \xrightarrow{p} T$ or $T_n \xrightarrow{\nu} T$ or $T_n \xrightarrow{n} T$. The following two questions arise naturally.

Question 1: If $\lambda_n \in \sigma(T_n)$ for $n \in \mathbb{N}$ and $\lambda_n \to \lambda$, must $\lambda \in \sigma(T)$?
Question 2: If $\lambda \in \sigma(T)$, is there $\lambda_n \in \sigma(T_n)$ for $n \in \mathbb{N}$ such that $\lambda_n \to \lambda$?

3.1 Answer to Question 1

If $T_n \xrightarrow{p} T$, then the answer to Question 1 is in the negative: In part (i) of Example 2.1, $T_n \xrightarrow{p} T$, $-1 \in \sigma(T_n)$ for each n, but $-1 \notin \sigma(T)$.

Modifications of this example show that multiplicities of the eigenvalues of T_n and T need not be the same. As in Example 2.1, let $X := \ell^2$. For $x := (x_1, x_2, \ldots) \in X$ and $n \in \mathbb{N}$, let $Tx := (x_1, 0, 0, \ldots)$ and $T_n x := (x_1, 0, \ldots, 0, x_n, 0, 0, \ldots)$. Then $T_n \xrightarrow{p} T$, 1 is an eigenvalue of T of multiplicity 1, while it is an eigenvalue of each T_n of multiplicity 2. On the other hand, if $Tx := (x_1, x_2, 0, 0, \ldots)$ and $T_n x := (x_1, x_2 - x_2/n, 0, 0, \ldots)$ for $x \in X$, then $T_n \xrightarrow{p} T$, 1 is an eigenvalue of T of multiplicity 2, while it is an eigenvalue of each T_n of multiplicity 1.

If $T_n \xrightarrow{\nu} T$, then the answer to Question 1 is in the affirmative. To show this, we prove a preliminary result whose proof depends on ideas we have already encountered in Propositions 2.2 and 2.5.

Lemma 3.1 *Let $T_n \xrightarrow{\nu} T$. Let E be a closed and bounded subset of $\mathbb{C}$ which is disjoint from $\sigma(T)$. Then E is disjoint from $\sigma(T_n)$ for all large n.*

Proof Since $T_n \xrightarrow{p} T$, there is $\alpha > 0$ such that $\|T_n\| \le \alpha$ for all $n \in \mathbb{N}$. Also, since E is closed and bounded, and the function $z \longmapsto \|R(z)\| \in \mathbb{R}$ is continuous on E, there is $\beta > 0$ such that $\|R(z)\| \le \beta$ for all $z \in E$.

Recall that $S_n(z) = (T_n - T)R(z)$ for $z \in \rho(T)$.

Assume first that $0 \in E$. For $z \in E$,

$$\begin{aligned}\|S_n(z)\| = \|(T_n - T)T\,T^{-1}R(z)\| &\le \|(T_n - T)T\|\,\|T^{-1}\|\,\|R(z)\| \\ &\le \|T^{-1}\|\beta\|(T_n - T)T\| \to 0.\end{aligned}$$

Hence for all large $n \in \mathbb{N}$ and all $z \in E$, $\|S_n(z)\| < 1$, and so by Proposition 2.2, $z \in \rho(T_n)$, that is, E is disjoint from $\sigma(T_n)$.

Next, assume that $0 \notin E$. Since E is closed, there is $\delta > 0$ such that $|z| \ge \delta$ for all $z \in E$. As in the proof of Proposition 2.5, for $z \in E$, we have

$$\begin{aligned}\|[S_n(z)]^2\| &= \frac{1}{|z|}\|\big((T_n - T)^2 + (T_n - T)TR(z)(T_n - T)\big)R(z)\| \\ &\le \frac{1}{\delta}\big(\|(T_n - T)^2\| + \|(T_n - T)T\|\beta(\alpha + \|T\|)\big)\beta \to 0.\end{aligned}$$

Hence for all large $n \in \mathbb{N}$ and all $z \in E$, $[\|S_n(z)]^2\| < 1$, and so by Proposition 2.5, $z \in \rho(T_n)$, that is, E is disjoint from $\sigma(T_n)$. □

Proposition 3.2 *Let $T_n \xrightarrow{\nu} T$. If $\lambda_n \in \sigma(T_n)$ and $\lambda_n \to \lambda$, then $\lambda \in \sigma(T)$.*

Proof Suppose $\lambda \notin \sigma(T)$. Since $\sigma(T)$ is a closed subset of $\mathbb{C}$, there is $r > 0$ such that $E := \{z \in \mathbb{C} : |z - \lambda| \le r\}$ is disjoint from $\sigma(T)$. By Lemma 3.1, E is disjoint from $\sigma(T_n)$ for all large n. Then $\lambda_n \in \sigma(T_n)$ and $\lambda_n \to \lambda$ is impossible. □

In particular, if $T_n \xrightarrow{n} T$, $\lambda_n \in \sigma(T_n)$ and $\lambda_n \to \lambda$, then $\lambda \in \sigma(T)$.

3.2 Answer to Question 2

The answer to Question 2 is in the negative even when $T_n \xrightarrow{n} T$.

Example 3.3 Let $X := \ell^2(\mathbb{Z})$, the space of all square summable "doubly infinite" complex sequences. For $x := (\ldots, x_{-2}, x_{-1}, x_0, x_1, x_2, \ldots)$, let $Tx := (\ldots, x_{-2}, x_{-1}, 0, x_1, x_2, \ldots)$ and $T_n x := (\ldots, x_{-2}, x_{-1}, x_0/n, x_1, x_2, \ldots)$. Then $\|T_n x - Tx\| = |x_0|/n$ for all $x \in X$, so that $\|T_n - T\| = 1/n \to 0$ as $n \to \infty$. However, it can be shown that $\sigma(T) = \{\lambda \in \mathbb{C} : |\lambda| \le 1\}$, whereas $\sigma(T_n) = \{\lambda \in \mathbb{C} : |\lambda| = 1\}$. (See Example 2.8 of [2]). Thus if $\lambda \in \mathbb{C}$ and $|\lambda| < 1$, then $\lambda \in \sigma(T)$, but there is no $\lambda_n \in \sigma(T_n)$ such that $\lambda_n \to \lambda$.

Faced with this negative conclusion, we consider some special points of $\sigma(T)$ about which positive results can be obtained.

Proposition 3.4 (i) *If λ is an isolated point of $\sigma(T)$ and $T_n \xrightarrow{n} T$, then there is $\lambda_n \in \sigma(T_n)$ such that $\lambda_n \to \lambda$.*
(ii) *If λ is a nonzero isolated point of $\sigma(T)$ and $T_n \xrightarrow{\nu} T$, then there is $\lambda_n \in \sigma(T_n)$ such that $\lambda_n \to \lambda$.*

The proof of this proposition involves the spectral projection

$$P(T, \Lambda) := -\frac{1}{2\pi i} \int_{\mathsf{C}} R(z)dz,$$

where Λ is a closed subset of $\sigma(T)$ such that $\sigma(T)\backslash\Lambda$ is also closed, and C is a Cauchy contour separating Λ from $\sigma(T)\backslash\Lambda$. (See Corollary 2.13 of [2]).

Further, if $\Lambda = \{\lambda\}$, C is a circle separating λ from $\sigma(T)\backslash\{\lambda\}$, and $P(T, \{\lambda\})$ is of finite rank m, then λ is an eigenvalue of T, and for all large n, there are m eigenvalues $\lambda_{n,1}, \ldots, \lambda_{n,m}$ of T_n (counted according to their algebraic multipicities) inside C, and their arithmetic mean $\widehat{\lambda}_n := (\lambda_{n,1} + \cdots + \lambda_{n,m})/m \to \lambda$. It is also possible to give error estimates for $|\widehat{\lambda}_n - \lambda|$. (See part (b) of Theorem 2.18 of [2]).

Remark 3.5 Let us comment on the special place of the complex number 0 in our discussion. If $T_n \xrightarrow{\nu} T$, then the case $z = 0$ in Proposition 2.5 and the case $\lambda = 0$ in Proposition 3.4 remain out of our reach.

If the Banach space X is infinite dimensional and if T is a compact operator, then it cannot be bounded below, and so $0 \notin \rho(T)$, that is, $0 \in \sigma(T)$. Compact operators, such as the integral operators considered in Sect. 2, are often approximated by operators whose range is finite dimensional. In the treatment of these operators, we need to avoid the complex number 0, as we shall see in the next section.

4 Finite Rank Operators

Let X be a linear space. A linear operator $\widetilde{T} : X \to X$ is said to be of finite rank if the range of $\widetilde{T}$ is finite dimensional. In this section, we shall give a canonical way of reducing the solution of the operator equation $zx - \widetilde{T}x = y$ and also of the eigenequation $\widetilde{T}x = \lambda x$, to matrix computations, provided $z \neq 0$ and $\lambda \neq 0$. In the classical literature, different ways were followed for this discretization depending on the nature of the finite rank operator. Whitley [12] suggested a canonical way applicable to all finite rank operators.

Let $\widetilde{T}$ be a finite rank operator on a linear space X. Then there are $x_1, \ldots, x_m \in X$ and linear functionals $f_1, \ldots, f_m$ on X such that

$$\widetilde{T}x := \sum_{i=1}^{m} f_i(x)x_i \quad \text{for } x \in X.$$

Neither the elements $x_1, \ldots, x_m$ nor the functionals $f_1, \ldots, f_m$ are assumed to be linearly independent. In this section, we shall assume that $\widetilde{T} : X \to X$ is presented to us as above.

Proposition 4.1 *Let X be a linear space, and $\widetilde{T} : X \to X$ be a linear operator of finite rank. Let A denote the $m \times m$ matrix having $f_i(x_j)$ as the entry in the ith row* *and the jth column. Let $0 \neq z \in \mathbb{C}$ and $y \in X$. Define $v := [f_1(y), \ldots, f_m(y)]^t$. Then for $x \in X$ and $u := [u_1, \ldots, u_m]^t \in \mathbb{C}^{m\times 1}$,*

$$zx - \widetilde{T}x = y \quad and \quad u = [f_1(x), \ldots, f_m(x)]^t$$

if and only if

$$zu - Au = v \quad and \quad x = \frac{1}{z}\Big(y + \sum_{j=1}^{m} u_j x_j\Big).$$

Proof Let $v_i := f_i(y)$ for $i = 1, \ldots, m$. We note that for $u := [u_1, \ldots, u_m]^t$,

$$zu - Au = v \Longleftrightarrow zu_i - \sum_{j=1}^{m} f_i(x_j)u_j = v_i \text{ for } i = 1, \ldots, m. \tag{1}$$

Assume that $x \in X$ and $zx - \widetilde{T}x = y$ and $u = [f_1(x), \ldots, f_m(x)]^t$. Then

$$zx = y + \sum_{j=1}^{m} f_j(x)x_j.$$

Applying f_i to the above equation, we obtain $zf_i(x) = f_i(y) + \sum_{j=1}^{m} f_j(x)f_i(x_j)$. Since $u_i = f_i(x)$ and $v_i = f_i(y)$ for $i = 1, \ldots, m$, we obtain $zu_i = v_i + \sum_{j=1}^{m} f_i(x_j)u_j$ for $i = 1, \ldots, m$, that is, $zu - Au = v$. Also, $x = (y + \sum_{j=1}^{m} u_j x_j)/z$.

Conversely, assume that $zu - Au = v$ and $x = (y + \sum_{j=1}^{m} u_j x_j)/z$. Then $\sum_{j=1}^{m} f_i(x_j)u_j = zu_i - v_i$ for $i = 1, \ldots, m$ by (1). Hence

$$\begin{aligned}\widetilde{T}x &= \sum_{i=1}^{m} f_i(x)x_i = \frac{1}{z}\sum_{i=1}^{m}\Big[f_i(y) + \sum_{j=1}^{m} u_j f_i(x_j)\Big]x_i \\ &= \frac{1}{z}\sum_{i=1}^{m}[v_i + zu_i - v_i]x_i = \sum_{i=1}^{m} u_i x_i \\ &= zx - y.\end{aligned}$$

Also, for $i = 1, \ldots, m$,

$$zu_i = v_i + \sum_{j=1}^{m} f_i(x_j)u_j = v_i + f_i\Big(\sum_{j=1}^{m} u_j x_j\Big) = f_i(y) + f_i(zx - y) = zf_i(x),$$

and so $u_i = f_i(x)$ since $z \neq 0$. This completes the proof. □

Thus to solve the operator equation $zx - \widetilde{T}x = y$, where $\widetilde{T}$ is of finite rank and $z \neq 0$, we may solve the matrix equation $zu - Au = v$, where $A := [f_i(x_j)]$ and $v := [f_1(y), \ldots, f_m(y)]^t$, and then let $x := \frac{1}{z}\Big(y + \sum_{j=1}^{m} u_j x_j\Big)$.

Corollary 4.2 *Any nonzero $\lambda \in \mathbb{C}$ is an eigenvalue of a finite rank operator $\widetilde{T}$ on a linear space X if and only if λ is an eigenvalue of the $m \times m$ matrix $A := [f_i(x_j)]$, and $\varphi \in X$ is an eigenvector of $\widetilde{T}$ corresponding to λ if and only if $\varphi = \left(\sum_{j=1}^m u_j x_j\right)/\lambda$, where $u := [u_1, \ldots, u_m]^t$ is an eigenvector of A corresponding to λ.*

Proof Let $y := 0$ in Proposition 4.1. Then $v := [f_1(0), \ldots, f_m(0)]^t = 0$. Writing λ for $z \in \mathbb{C}$, and φ for $x \in X$, we see that $\lambda\varphi - \widetilde{T}\varphi = 0$ and $u_i = f_i(\varphi)$ for $i = 1, \ldots, m$ if and only if $\lambda u - Au = 0$ and $\varphi = \left(\sum_{j=1}^m u_j x_j\right)/\lambda$. Further, if $u = 0$, that is, $u_j = 0$ for all $j = 1, \ldots, m$, then clearly $\varphi = 0$. Also, if $\varphi = 0$, then $u_i = f_i(\varphi) = 0$ for $i = 1, \ldots, m$, that is, $u = 0$. Thus φ is an eigenvector of $\widetilde{T}$ corresponding to λ if and only if u is an eigenvector of A corresponding to λ. □

Thus to solve the operator eigenequation $\widetilde{T}\varphi = \lambda\varphi$, where $\widetilde{T}$ is of finite rank and $\lambda \neq 0$, we may solve the matrix eigenequation $Au = \lambda u$, where $A := [f_i(x_j)]$, and then let $\varphi := \left(\sum_{j=1}^m u_j x_j\right)/\lambda$, where $u := [u_1, \ldots, u_m]^t$.

The finite rank operators such as $T_n^D, T_n^P, T_n^S, T_n^G, T_n^N$ appearing in most classical methods of approximation are always presented to us in the form $\sum_{j=1}^m f_j(x)x_j$, $x \in X$, considered here. This will be apparent from the following typical examples.

Example 4.3 Let $X := C([a, b])$.
(i) For $x \in X$, let

$$(T_n^D x)(s) := \int_a^b \sum_{j=1}^{n} x_{n,j}(s)\, y_{n,j}(t)x(t)dt, \quad s \in [a, b].$$

Then $T_n^D x = \sum_{j=1}^{n} f_j(x)x_j$, where $x_j := x_{n,j}$, and $f_j(x) := \int_a^b y_{n,j}(t)x(t)dt$ for $x \in X,\ j = 1, \ldots, n$.

(ii) For $x \in X$, let

$$T_n^P x := \sum_{j=1}^{n} (Tx)(t_{n,j})e_{n,j}.$$

Then $T_n^P x = \sum_{j=1}^{n} f_j(x)x_j$, where $x_j := e_{n,j}$, and $f_j(x) := (Tx)(t_{n,j})$ for $x \in X,\ j = 1, \ldots, n$.

(iii) For $x \in X$, let

$$(T_n^N x)(s) := \sum_{j=1}^{n} w_{n,j} k(s, t_{n,j}) x(t_{n,j}), \quad s \in [a, b].$$

Then $T_n^N x = \sum_{j=1}^{n} f_j(x) x_j$, where $x_j(s) := k(s, t_{n,j})$ for $s \in [a, b]$, and $f_j(x) := w_{n,j} x(t_{n,j})$ for $x \in X, \ j = 1, \ldots, n$.

We conclude this article by considering two integral operators with apparently similar kernels. Let $[a, b] := [0, 1]$. Define $h(s, t) := e^{s+t}$ and $k(s, t) := e^{st}$ for $s, t \in [0, 1]$. For $x \in C([0, 1])$, let

$$(\widetilde{T} x)(s) := \int_0^1 h(s, t) x(t) dt \quad \text{and} \quad (T x)(s) := \int_0^1 k(s, t) x(t) dt, \quad s \in [0, 1].$$

Then $\widetilde{T}$ is a finite rank operator. In fact, for $x \in C([0, 1])$, $\widetilde{T} x = f_1(x) x_1$, where $x_1(s) := e^s$ for $s \in [0, 1]$, and $f_1(x) := \int_0^1 e^t x(t) dt$ for $x \in X$. On the other hand, T is not of finite rank. Noting that $e^{st} = \sum_{j=0}^{\infty} \frac{s^j t^j}{j!}$ for $s, t \in [0, 1]$, where the series converges uniformly on $[0, 1] \times [0, 1]$, we may let $k_n(s, t) := \sum_{j=0}^{n} \frac{s^j t^j}{j!}$ for $s, t \in [0, 1]$. Then $\|k(\cdot, \cdot) - k_n(\cdot, \cdot)\|_\infty \to 0$ as $n \to \infty$. For $x \in C([0, 1])$, define

$$(T_n^D x)(s) := \int_0^1 k_n(s, t) x(t) dt, \quad s \in [0, 1].$$

Then $T_n^D \xrightarrow{n} T$. In fact, for each fixed $n \in \mathbb{N}$, $(T_n^D)(x) = \sum_{j=0}^{n} f_j(x) x_j$, where $x_j(s) := \frac{s^j}{j!}$ for $s \in [0, 1]$, and $f_j(x) := \int_0^1 t^j x(t) dt$ for $x \in C([0, 1])$, $j = 1, \ldots, n$.

References

1. Ahues, M.: A class of strongly stable operator approximations. J. Austral. Math. Soc. Ser. B **28**, 435–442 (1987)
2. Ahues, M., Largillier, A., Limaye, B.V.: Spectral Computations for Bounded Operators. Chapman and Hall/CRC, Boca Raton (2001)

3. Anselone, P.M.: Collectively Compact Operator Approximation Theory and Applications to Integral Equations. Prentice-Hall, Englewood Cliffs, N.J (1971)
4. Anselone, P.M., Moore, R.: Approximate solution of integral and operator equations. J. Math. Anal. Appl. **9**, 268–277 (1964)
5. Atkinson, K.E.: The numerical solution of eigenvalue problem for compact integral operators. Trans. Amer. Math. Soc. **129**, 458–465 (1967)
6. Atkinson, K.E.: The Numerical Solution of Integral Equations of the Second Kind. Cambridge University Press, Cambridge (1997)
7. Atkinson, K.E.: A Personal Perspective on the History of the Numerical Analysis of Fredholm Integral Equations of the Second Kind. The Birth of Numerical Analysis, World Scientific, Singapore (2010)
8. Bouldin, R.: Operator approximations with stable eigenvalues. J. Austral. Math. Soc. Ser. A **49**, 250–257 (1990)
9. Kulkarni, R.P.: A new superconvergent projection method for approximate solutions of eigenvalue problems. Numer. Funct. Anal. Optimiz. **24**, 75–84 (2003)
10. Nair, M.T.: On strongly stable approximations, J. Austral. Math. Soc. Ser. A **52**, 251–260 (1992)
11. Nair, M.T.: Linear Operator Equations: Approximation and Regularization. World Scientific, Singapore (2009)
12. Whitley, R.: The stability of finite rank methods with applications to integral equations, SIAM. J. Numer. Anal. **23**, 118–134 (1986)

The Nullity Theorem, Its Generalization and Applications

S.H. Kulkarni

Abstract The Nullity theorem says that certain pairs of submatrices of a square invertible matrix and its inverse (known as complementary submatrices) have the same nullity. Though this theorem has been around for quite some time and also has found several applications, some how it is not that widely known. We give a brief account of the Nullity Theorem, consider its generalization to infinite dimensional spaces, called the Null Space Theorem and discuss some applications.

Keywords Nullity theorem · Null space theorem · Tridiagonal operator · Rank · Generalized inverse

Mathematics Subject Classification (2000) 15A09 · 47A05

1 Introduction

This article is based on the talks given by me at several places including the one at the International Conference on Semigroups, Algebras and Operator Theory (ICSAOT)-2014 held at the Department of Mathematics, Cochin University of Science and Technology. It has two objectives. The first is to make the Nullity Theorem known more widely. I myself started taking interest in this topic after listening to a talk by Prof. R.B. Bapat in a workshop on Numerical Linear Algebra at I. I. T. Guwahati. The second is to consider its generalization to infinite dimensional spaces and some applications of this generalization. We begin with some motivation for the Nullity Theorem.

Recall that a square matrix $A = [\alpha_{ij}]$ of order n is called *tridiagonal* if

$$\alpha_{ij} = 0 \quad \text{for} \quad |i - j| > 1$$

S.H. Kulkarni (✉)
Department of Mathematics, Indian Institute of Technology - Madras, Chennai 600036, India
e-mail: shk@iitm.ac.in

P.G. Romeo et al. (eds.), *Semigroups, Algebras and Operator Theory*,
Springer Proceedings in Mathematics & Statistics 142,
DOI 10.1007/978-81-322-2488-4_12

It is well known that the tridiagonal matrices form a very useful class of matrices with applications in many areas. Tridiagonal matrices have a close relationship with the second-order linear differential equations. (See [8] for more discussion on this topic.) In particular, the final step in solving a second-order linear differential equation numerically by finite difference method involves solving a system of linear equations whose coefficient matrix is tridiagonal. Also the transition probability matrix in a Birth and Death Process happens to be a tridiagonal matrix.

Such a tridiagonal matrix is described completely by $3n - 2$ numbers (n on the main diagonal and $n - 1$ on each of superdiagonal and subdiagonal). In general, if a tridiagonal matrix is invertible, its inverse need not be tridiagonal. It is very easy to construct examples to illustrate this. However, we may still expect that the inverse can be described completely by $3n - 2$ parameters. This is indeed true. It is known that if A is a tridiagonal matrix of order n and if A is also invertible, then every submatrix of A^{-1} that lies on or above the main diagonal is of rank ≤ 1. Similar statement is true of submatrices lying on or below the main diagonal. This result is known at least since 1979. (See [2]) Several proofs of this result are available in the literature. The article [8] contains some of these proofs, references to these and other proofs, and also a brief history and comments about possible generalizations.

In view of this result, the inverse can be described using $3n - 2$ parameters as follows: To start with we can choose $4n$ numbers $a_j, b_j, c_j, d_j, \quad j = 1, \ldots n$ such that

$$\begin{aligned}(A^{-1})_{ij} &= a_i b_j \quad \text{for} \quad i \leq j \quad \text{and} \\ &= c_i d_j \quad \text{for} \quad j \leq i\end{aligned}$$

These $4n$ numbers have to satisfy following constraints.

$$a_i b_i = c_i d_i \quad \text{for} \quad i = 1, \ldots, n \quad \text{and} \quad a_1 = 1 = c_1$$

This result has been used to construct fast algorithms to compute the inverse of a tridiagonal matrix or to find solutions of a linear system whose coefficient matrix is tridiagonal. (See [8]) One proof of this theorem depends on the Nullity Theorem. This theorem uses the idea of complementary submatrices. Let A and B be square matrices of order n. Suppose M is a submatrix of A and N is a submatrix of B. We say that M and N are *complementary* if row numbers not used in one are the column numbers used in the other. More precisely, let I and J be subsets of the set $\{1, 2, \ldots, n\}$ and let I^c denote the complement of I. Let $A(I, J)$ denote the submatrix of A obtained by choosing rows in I and columns in J. Then $A(I, J)$ and $B(J^c, I^c)$ are complementary submatrices. With this terminology, the Nullity Theorem has a very simple formulation.

Theorem 1.1 (Nullity Theorem) *Complementary submatrices of a square matrix and its inverse have the same nullity.*

As an illustration we can consider the following. Suppose $k < n$ and a square matrix M of order n is partitioned into submatrices as follows:

$$M = \begin{bmatrix} A_k & B_k \\ C_k & D_k \end{bmatrix}$$

Here A_k is the submatrix obtained from A by choosing the first k rows and the first k columns. Assume that M is invertible and its inverse is partitioned similarly as follows:

$$M^{-1} = \begin{bmatrix} P_k & Q_k \\ R_k & S_k \end{bmatrix}$$

Then the Nullity Theorem says that

$$\text{nullity}(A_k) = \text{nullity}(S_k), \quad \text{nullity}(D_k) = \text{nullity}(P_k)$$

$$\text{nullity}(B_k) = \text{nullity}(Q_k), \quad \text{nullity}(C_k) = \text{nullity}(R_k)$$

This Nullity Theorem has been in the literature for quite some time (at least since 1984), but it does not seem to be that widely well known. In [8], Gilbert Strang and Tri Ngyuen have given an account of this Nullity Theorem. They have given a proof of this theorem and discussed its consequences for ranks of some submatrices. In particular, they prove a very interesting fact that the submatrices of a banded invertible matrix lying above or below the main diagonal have low ranks. While discussing literature and alternative proofs, the authors make the following remark.

"A key question will be the generalization to infinite dimensions."

We give a brief account of such a generalization called the "Null Space Theorem". This was also reported in [5].

We recall a few standard notations, definitions, and results that are used to prove the main result. For vector spaces X, Y, we denote by $L(X, Y)$ the set of all linear operators from X to Y. For an operator $T \in L(X, Y)$, $N(T)$ denotes the null space of T and $R(T)$ denotes the range of T. Thus $N(T) := \{x \in X : T(x) = 0\}$ and $R(T) := \{T(x) : x \in X\}$.

As usual, $L(X, X)$ will be denoted by $L(X)$. A map $P \in L(X)$ is called a *projection* if $P^2 = P$. Let $P \in L(X)$ and $Q \in L(Y)$ be projections. The restriction of QTP to $R(P)$ can be viewed as a linear operator from $R(P)$ to $R(Q)$. This is called a *section of* T *by* P *and* Q and will be denoted by $T_{P,Q}$. It is called a *finite section*, if $R(P)$ and $R(Q)$ are finite dimensional. When T is invertible, the section $T^{-1}_{I_Y-Q, I_X-P}$ of T^{-1} is called the *complementary section* of $T_{P,Q}$. With this terminology, our Null Space Theorem can be stated in the following very simple form.

There is a linear bijection between the null spaces of the complementary sections of T and T^{-1} (Theorem 2.1).

Its proof is also very simple. It is given in the next section. When X and Y are finite dimensional, T is represented by a matrix and complementary submatrices correspond to complementary sections. (See [8]) Thus there is a linear bijection between the null spaces of the complementary submatrices of T and T^{-1}. Hence they have the same nullity. This is the Nullity Theorem (Theorem 1.1).

The authors of [8] have discussed several applications of the Nullity Theorem. For example, if T is an invertible tridiagonal matrix, then every submatrix of T^{-1} that lies on or above the main diagonal or on and below the main diagonal is of rank ≤ 1. However, proofs of these applications involve the famous Rank–Nullity Theorem (called the "Fundamental Theorem of Linear Algebra" in [7]) apart from the Nullity Theorem. Hence a straightforward imitation of these proofs to infinite dimensional case may or may not work, though the results may very well be true. Such an approach may work when the sections can be viewed as operators on finite dimensional spaces. In general, we need a different approach. This is attempted in the third section. We prove that if a tridiagonal operator on a Banach space with a Schauder basis is invertible, then certain sections of T^{-1} are of rank ≤ 1. (Theorem 3.1) This is followed by some illustrative examples and remarks about possible extensions.

2 Main Result

Theorem 2.1 (The Null Space Theorem) *Let X, Y be vector spaces, $P \in L(X)$, $Q \in L(Y)$ be projections and $T \in L(X, Y)$ be invertible. Then there is a linear bijection between the null space of the section $T_{P,Q}$ of T and the null space of its complementary section $T^{-1}_{I_Y-Q,I_X-P}$ of T^{-1}.*

Proof (*See also* [5]) Let $x \in N(T_{P,Q})$. This means that $x \in R(P)$ so that $P(x) = x$ and $QTP(x) = 0$. Hence $QT(x) = 0$, that is, $(I_Y - Q)T(x) = T(x)$. Thus $T(x) \in R(I_Y - Q)$. Also, $(I_X - P)T^{-1}(I_Y - Q)T(x) = (I_X - P)T^{-1}T(x) = (I_X - P)(x) = 0$. Hence $T(x) \in N((I_X - P)T^{-1}(I_Y - Q))$. This means $T(x) \in N(T^{-1}_{I_Y-Q,I_X-P})$. This shows that the restriction of T to $N(T_{P,Q})$ maps $N(T_{P,Q})$ into $N((I_X - P)T^{-1}(I_Y - Q))$. Since T is invertible, this map is already injective. It only remains to show that it is onto. For this let $y \in N(T^{-1}_{I_Y-Q,I_X-P})$. This means $y \in R(I_Y - Q)$ and $(I_X - P)T^{-1}(I_Y - Q)(y) = 0$. We shall show that $T^{-1}(y) \in N(T_{P,Q})$. Since $y \in R(I_Y - Q)$, we have $(I_Y - Q)(y) = y$. Thus $Q(y) = 0$. Next, $0 = (I_X - P)T^{-1}(I_Y - Q)(y) = (I_X - P)T^{-1}(y)$. This implies that $PT^{-1}(y) = T^{-1}(y)$, that is , $T^{-1}(y) \in R(P)$. Also $QTPT^{-1}(y) = QTT^{-1}(y) = Q(y) = 0$. Thus $T^{-1}(y) \in N(QTP)$. Hence $T^{-1}(y) \in N(T_{P,Q})$. □

Remark 2.2 As pointed out in the Introduction, this Null Space Theorem implies the Nullity Theorem (Theorem 1.1).

3 Ranks of Submatirces

While considering infinite matrices, the products involve infinite sums, leading naturally to the questions of convergence. Hence it is natural to consider these questions in the setting of a Banach space X with a Schauder basis $A = \{a_1, a_2, \ldots\}$. We refer to [4, 6] for elementary concepts in Functional Analysis.

Theorem 3.1 *Let X be a Banach space with a Schauder basis $A = \{a_1, a_2, \ldots\}$. Let T be a bounded (continuous) linear operator on X. Suppose the matrix of T with respect to A is tridiagonal. If T is invertible, then every submatrix of the matrix of T^{-1} with respect to A that lies on or above the main diagonal (or on or below the main diagonal) is of rank ≤ 1.*

Proof Let M be the matrix of T with respect to A. Then M is infinite matrix of the form

$$M = \begin{bmatrix} \delta_1 & \alpha_1 & 0 & 0 & . \\ \beta_2 & \delta_2 & \alpha_2 & 0 & . \\ 0 & \beta_3 & \delta_3 & \alpha_3 & . \\ 0 & 0 & . & . & . \\ . & . & . & . & . \end{bmatrix}$$

Then the matrix of T^{-1} with respect to A is M^{-1}. Let $M^{-1} = [\gamma_{i,j}]$. Let C_j denote the j-th column of M^{-1} and R_i denote the i-th row of M^{-1}. Thus

$$M^{-1} = \left[C_1\ C_2 \ldots\right] = \begin{bmatrix} R_1 \\ R_2 \\ . \\ . \\ . \end{bmatrix}$$

Further for a fixed natural number k, let C_j^k denote the column vector obtained by deleting the first $k-1$ entries from C_j. Thus

$$C_j^k = \begin{bmatrix} \gamma_{k,j} \\ \gamma_{k+1,j} \\ . \\ . \\ . \end{bmatrix}$$

Similarly, let R_i^k denote the row vector obtained by deleting first $k-1$ entries from R_i.

Next let P_k denote the submatrix of M^{-1} given by

$$P_k = [\gamma_{i,j}, i \geq k, 1 \leq j \leq k] = \left[C_1^k\ C_2^k\ ..\ C_k^k\right]$$

Similarly, let Q_k denote the submatrix of M^{-1} given by

$$Q_k = [\gamma_{i,j}, 1 \leq i \leq k, j \geq k] = \begin{bmatrix} R_1^k \\ R_2^k \\ \cdot \\ \cdot \\ R_k^k \end{bmatrix}$$

Note that every submatrix of M^{-1} that lies on or above the main diagonal is a submatrix of Q_k for some k and every submatrix of M^{-1} that lies on or below the main diagonal is a submatrix of P_k for some k. Thus it is sufficient to show that P_k and Q_k are of rank ≤ 1 for each k.

We shall give two proofs of this assertion.

First proof:

The assertion is evident for $k = 1$.

Now consider the equation $M^{-1}M = I$, that is,

$$\left[C_1\ C_2 \ldots\right] \begin{bmatrix} \delta_1 & \alpha_1 & 0 & 0 & . \\ \beta_2 & \delta_2 & \alpha_2 & 0 & . \\ 0 & \beta_3 & \delta_3 & \alpha_3 & . \\ 0 & 0 & . & . & . \\ . & . & . & . & . \end{bmatrix} = \left[e_1\ e_2 \ldots\right]$$

where, as usual, e_j denotes the column matrix whose j-th entry is 1 and all other entries are 0.

Equating the first columns on both sides of the above equation, we get

$$\delta_1 C_1 + \beta_2 C_2 = e_1$$

This, in particular, implies that at least one of δ_1, β_2 is not zero.

Deleting the first entries from all the column vectors in the above equation, we get

$$\delta_1 C_1^2 + \beta_2 C_2^2 = e_1^2 = 0$$

This shows that $\{C_1^2, C_2^2\}$ is a linearly dependent set, that is the matrix

$$P_2 = \left[C_1^2\ C_2^2\right]$$

is of rank ≤ 1.

Next we equate the second column on both sides of the equation. Then

$$\alpha_1 C_1 + \delta_2 C_2 + \beta_3 C_3 = e_2$$

Hence one of $\alpha_1,\ \delta_2,\ \beta_3$ is not zero.

Now deleting the first two entries from all the vectors appearing in this equation, we get

$$\alpha_1 C_1^3 + \delta_2 C_2^3 + \beta_3 C_3^3 = e_2^3 = 0$$

Since $\{C_1^2, C_2^2\}$ is a linearly dependent set, one of the vectors, say C_2^2 is a scalar multiple of the other, that is, C_1^2. This implies that C_2^3 is a scalar multiple of C_1^3. Now the above equation shows that C_3^3 is also a scalar multiple of C_1^3. Hence

$$P_3 = \left[C_1^3 \; C_2^3 \; C_3^3\right]$$

is of rank ≤ 1.

Proceeding in this way (more precisely, by Mathematical Induction), we can show that

P_k is of rank ≤ 1 for each k.

Following essentially the same technique, equating the rows of both sides of the equation $MM^{-1} = I$, we can show that Q_k is of rank ≤ 1 for each k.

This completes the first proof.

Second proof:

Recall that since $A = \{a_1, a_2, \ldots\}$ is a Schauder basis of X, every $x \in X$ can be expressed uniquely as $x = \sum_{j=1}^{\infty} \alpha_j a_j$ for some scalars α_j. Let X_n denote the linear span of $A_n := \{a_1, a_2, \ldots a_n\}$ and define a map $\pi_n : X \to X$ by $\pi_n(x) = \sum_{j=1}^{n} \alpha_j a_j$. Then π_n is a projection with $R(\pi_n) = X_n$. Also note that for each k, P_k as defined above is the matrix of the section $T^{-1}_{\pi_k, I-\pi_{k-1}}$ of the operator T^{-1}. As noted earlier, this can be viewed as an operator on $R(\pi_k) = X_k$. By the Null Space Theorem (Theorem 2.1), there is a linear bijection between the null space of this section and its complementary section, that is, the section $T_{\pi_{k-1}, I-\pi_k}$ of the operator T. This can be viewed as an operator on $R(\pi_{k-1}) = X_{k-1}$. It can be seen (in many ways) that this is in fact the zero operator on X_{k-1}. (The matrix of this section is the submatrix of M obtained by choosing the first $k-1$ columns and not choosing the first k rows. This is a zero matrix because M is tridiagonal.) Thus the null space of the section $T_{\pi_{k-1}, I-\pi_k}$ coincides with X_{k-1}. Hence the null space of the complementary section $T^{-1}_{\pi_k, I-\pi_{k-1}}$ is also of dimension $k-1$. This implies that its rank is 1 as it is an operator on X_k.

Thus P_k is of rank 1 for each k.

In a similar way, we can show that Q_k is of rank 1 for each k.

This completes the second proof □

Example 3.2 Let ℓ^2 denote the Hilbert space of square summable sequences and let $E = \{e_1, e_2, \ldots\}$ be the orthonormal basis, where as usual e_j denotes the sequence whose j-th entry is 1 and all other entries are 0. Let R denote the Right Shift operator given by

$$R(x) = (0, x(1), x(2), \ldots), \quad x \in \ell^2.$$

Consider a complex number c with $|c| < 1$ and let $T = I - cR$. Then the matrix of T with respect to the orthonormal basis E is tridiagonal and is given by

$$\begin{bmatrix} 1 & 0 & 0 & . & . \\ -c & 1 & 0 & 0 & . \\ 0 & -c & 1 & 0 & . \\ . & . & . & . & . \\ . & . & . & . & . \end{bmatrix}$$

It can be easily checked that T is invertible and

$$T^{-1} = \sum_{j=0}^{\infty} c^j R^j$$

Thus the matrix of T^{-1} with respect to the orthonormal basis E is given by

$$\begin{bmatrix} 1 & 0 & 0 & . & . \\ c & 1 & 0 & 0 & . \\ c^2 & c & 1 & 0 & . \\ . & . & . & . & . \\ . & . & . & . & . \end{bmatrix}$$

It is easily seen that every submatrix of the above matrix of T^{-1} that lies on or above (or on or below) the main diagonal is of rank 0 or 1.

Remark 3.3 In view of the above Theorem, T^{-1} or equivalently, $M^{-1} = [\gamma_{i,j}]$ can be described completely by using four sequences $\{a_n\}$, $\{b_n\}$, $\{c_n\}$, $\{d_n\}$ as follows: $\gamma_{i,j} = a_i b_j$ for $j \geq i$ and $\gamma_{i,j} = c_i d_j$ for $j \leq i$. Also, since for $i = j$, $\gamma_{i,i} = a_i b_i = c_i d_i$, these are essentially only three sequences. This should be expected as the tridiagonal operator T (matrix M) is completely described by three sequences, namely, $\{\alpha_n\}$, $\{\beta_n\}$, $\{\delta_n\}$. This can be useful in devising fast methods of computing T^{-1}. (See the Introduction of [8].)

Remark 3.4 It is also easy to see that the above proof can be easily modified in an obvious manner to a natural generalization that allows the matrix M of T to have a wider band. Suppose $M = [m_{i,j}]$ is such that $m_{i,j} = 0$ for $|i - j| > p$. (Thus $p = 1$ corresponds to tridiagonal operator.) Then using the same method, we can prove the following: every submatrix of the matrix of M^{-1} that lies above the pth subdiagonal or below the pth superdiagonal is of rank $\leq p$.

A careful look at the proof of Theorem 3.1 in fact shows that we have actually proved a more general result.

Theorem 3.5 *Let X be a Banach space with a Schauder basis $A = \{a_1, a_2, \ldots\}$. Let T be a bounded (continuous) linear operator on X. Suppose the matrix of T with respect to A is tridiagonal. If T has a bounded left inverse S, then every submatrix of the matrix of S with respect to A that lies on or below the main diagonal is of rank ≤ 1. Similarly, if T has a bounded right inverse U, then every submatrix of the matrix of U with respect to A that lies on or above the main diagonal is of rank ≤ 1.*

Remark 3.6 As a simple example of the above Theorem 3.5, we may again consider the right shift operator R on ℓ^2 discussed in Example 3.2. Let L denote the Left Shift operator given by

$$L(x) = (x(2), x(3), \ldots), \quad x \in \ell^2.$$

Then L is a left inverse of R. Clearly, every submatrix of the matrix of L with respect to A that lies on or below the main diagonal is of rank ≤ 1.

Remark 3.7 Since left (right) inverse is one among the family of generalized inverses, Theorem 3.5 also raises an obvious question: Is there an analogue of Theorem 3.5 for other generalized inverses, in particular for Moore–Penrose pseudoinverse? Such results are known for matrices. (See [1]) Information on generalized inverses of various types can be found in [3].

Remark 3.8 In order to draw significant conclusions in the context of differential equations, we need an extension of Theorem 3.1 for unbounded operators.

Acknowledgments The author thanks Prof. Gilbert Strang and Prof. R.B. Bapat for several useful discussions.

References

1. Bapat, R.B.: On generalized inverses of banded matrices. Electron. J. Linear Algebra **16**, 284–290 (2007) (MR2349923 (2008k:15004))
2. Barrett, W.W.: A theorem on inverses of tridiagonal matrices. Linear Algebra Appl. **27**, 211–217 (1979) (MR0545734 (80i:15003))
3. Ben-Israel, A., Greville, T.N.E.: Generalized Inverses, 3rd edn. Springer, New York (2003)
4. Bollobás, B.: Linear Analysis. Cambridge University Press, Cambridge (1990) (MR1087297 (92a:46001))
5. Kulkarni, S.H.: The null space theorem. Linear Algebra Appl. (2015). doi:10.1016/j.laa.2015.01.030
6. Limaye, B.V.: Functional Analysis, 2nd edn. New Age, New Delhi (1996) (MR1427262 (97k:46001))
7. Strang, G.: Linear Algebra and Its Applications, 2nd edn. Academic Press, New York (1980) (MR0575349 (81f:15003))
8. Strang, G., Nguyen, T.: The interplay of ranks of submatrices. SIAM Rev. **46**(4), 637–646 (2004) ((electronic). MR2124679 (2005m:15015))

Role of Hilbert Scales in Regularization Theory

M.T. Nair

Abstract Hilbert scales, which are generalizations of Sobolev scales, play crucial roles in the regularization theory. In this paper, it is intended to discuss some important properties of Hilbert scales with illustrations through examples constructed using the concept of Gelfand triples, and using them to describe source conditions and for deriving error estimates in the regularized solutions of ill-posed operator equations. We discuss the above with special emphasis on some of the recent work of the author.

Keywords Hilbert scales · Sobolev scales · Ill-posed equations · Regularization · Tikhonov regularization · Source sets · Discrepancy principle · Order optimal gelfand triple · Ill-posed problem · Gelfand triple

AMS Subject Classification 65R10 · 65J10 · 65J20 · 65R20 · 45B05 · 45L10 · 47A50

1 What are Hilbert Scales?

Definition 1.1 A family of Hilbert spaces H_s, $s \in \mathbb{R}$, is called a **Hilbert scale** if $H_t \subseteq H_s$ whenever $t > s$ and the inclusion is a continuous embedding, i.e., there exists $c_{s,t} > 0$ such that

$$\|x\|_s \leq c_{s,t}\|x\|_t \quad \forall x \in H_t. \qquad \square$$

Examples of Hilbert scales are constructed by first defining H_s for $s \geq 0$, and then defining H_s for $s < 0$ using the concept of a **Gelfand triple**. So, let us consider the definition and properties of Gelfand triples.

Throughout the paper, we shall consider the scalar field is $\mathbb{K}$ which is either the field $\mathbb{R}$ of real numbers of the field $\mathbb{C}$ of complex numbers.

M.T. Nair (✉)
Department of Mathematics, IIT Madras, Chennai 600036, India
e-mail: mtnair@iitm.ac.in

P.G. Romeo et al. (eds.), *Semigroups, Algebras and Operator Theory*,
Springer Proceedings in Mathematics & Statistics 142,
DOI 10.1007/978-81-322-2488-4_13

1.1 Gelfand Triple

Let V be a dense subspace of a Hilbert space H with norm $\|\cdot\|$. Suppose V is also a Hilbert space with respect to a norm $\|\cdot\|_V$ such that the inclusion of V into H is continuous, i.e., there exists $c > 0$ such that

$$\|x\| \le c\|x\|_V \quad \forall x \in V.$$

For $x \in H$, let

$$\|x\|_* := \sup\{|\langle u, x\rangle| : v \in V,\ \|v\|_V \le 1\}.$$

It can be easily seen that $\|\cdot\|_*$ is a norm on H. In fact, it is weaker than the original norm $\|\cdot\|$. Indeed, for $x \in H,\ v \in V$

$$|\langle x, v\rangle| \le \|x\|\,\|v\| \le c\|x\|\,\|v\|_V$$

so that

$$\|x\|_* \le \|x\| \quad \forall u \in H.$$

Let $\widetilde{V}$ be the completion of H with respect to the norm $\|\cdot\|_*$.

Definition 1.2 The triple $(V, H, \widetilde{V})$ is called a **Gelfand triple**. □

We show that $\widetilde{V}$ is linearly isometric with V', the dual of V of all continuous linear functionals on V. For this purpose, for each $x \in H$, consider the the map $f_x : V \to \mathbb{K}$ defined by

$$f_x(v) = \langle v, x\rangle, \quad v \in V.$$

Then f_x is linear and

$$|f_v(u)| = |\langle u, v\rangle| \le \|u\|\,\|v\| \le c\|u\|_1\|v\|, \quad u \in H_1.$$

Hence,

$$f_x \in V', \quad \|f_x\| = \|x\|_*$$

and the map $x \mapsto f_x$ is a linear isometry from H into $\widetilde{V}$. Further, we have the following.

Theorem 1.3 *The subspace $\{f_x : x \in H\}$ of V' is dense in V'.*

Proof By Hahn–Banach extension theorem, it is enough to prove that for $\varphi \in V''$, $\varphi(f_x) = 0$ for all $x \in H$ implies $x = 0$. So, let $\varphi \in V''$ such that $\varphi(f_x) = 0$ for all $x \in H$. Since V is reflexive, there exists $u \in V$ such that

$$\varphi(f) = f(u) \quad \forall f \in V'.$$

Thus, $f_x(u) = 0$ for all $x \in H$, i.e., $\langle u, x\rangle = 0$ for all $x \in H$. Hence, $x = 0$. □

As a consequence of the above theorem and the remarks preceding the theorem, we obtain the following.

Theorem 1.4 *The space* $\widetilde{V}$ *is linearly isometric with* V',

$$V \subseteq H \subseteq \widetilde{V}$$

and the inclusions are continuous embeddings.

2 Examples of Hilbert Scales

Example 2.1 Let H be a Hilbert space and $L : D(L) \subseteq H \to H$ be a densely defined strictly positive self adjont operator which is also coercive, i.e.,

$$\langle Lx, x\rangle \geq \gamma \|x\|^2 \quad \forall x \in H$$

for some $\gamma > 0$. Consider the dense subspace $X = \cap_{k=1}^{\infty} D(L^k)$, and for $s > 0$, let

$$\langle u, v\rangle_s := \langle L^s u, L^s v\rangle, \quad u \in X.$$

$$\|u\|_s := \|L^s u\|, \quad u \in X.$$

Then $\langle \cdot, \cdot\rangle_s$ is an inner product on X with the corresponding norm

$$\|u\|_s := \|L^s u\|, \quad u \in X.$$

Here, for $s \in \mathbb{R}$, the operator L^s is defined via spectral theorem, i.e.,

$$L^s = \int_0^{\infty} \lambda^s dE_\lambda,$$

where $\{E_\lambda\}_{\lambda \in \mathbb{R}}$ is the resolution of identity associated with the operator L.

Let H_s be the completion of X with respect to $\|\cdot\|_s$. It can be seen that H_s is a dense subspace of H as a vector space. Also, since L is bounded below, L^s is also bounded below, so that there exists $c_s > 0$ such that

$$\|L^s u\| \geq c_s \|u\| \quad \forall u \in X.$$

Thus, H_s is continuously embedded in H. Thus, we have the Gelfand triple (H_s, H, H_{-s}) with $H_{-s} := \widetilde{H}_s$. Note that

$$s \leq t \implies H_t \subseteq H_s$$

and the inclusion is continuous, with $H_0 = H$. Thus, the family $\{H_s : s \in \mathbb{R}\}$ is a Hilbert scale, called the *Hilbert scale generated by L*. □

Example 2.2 Let H be a separable Hilbert space and $\{u_n : n \in \mathbb{N}\}$ be an orthonormal basis of H. Let (σ_n) be a sequence of positive real numbers with $\sigma_n \to 0$. For $s \geq 0$, let

$$H_s := \{x \in H : \sum_{n=1}^{\infty} \frac{|\langle x, u_n\rangle|^2}{\sigma_n^{2s}} < \infty\}.$$

Then H_s is a Hilbert space with inner product

$$\langle x, y\rangle_s := \sum_{n=1}^{\infty} \frac{\langle x, u_n\rangle\langle u_n, y\rangle}{\sigma_n^{2s}}.$$

The corresponding norm $\|\cdot\|_s$ is given by

$$\|x\|_s^2 := \sum_{n=1}^{\infty} \frac{|\langle x, u_n\rangle|^2}{\sigma_n^{2s}}.$$

We may observe that

$$\|x\| \leq \|x\|_s \quad \forall x \in H_s,\ s > 0.$$

Thus, (H_s, H, H_{-s}), with $H_0 = H$ and $H_{-s} := \widetilde{H}_s$, is a Gelfand triple for each $s > 0$, and $\{H_s : s \in \mathbb{R}\}$ is a Hilbert scale.

Let $T : H \to H$ be defined by

$$Tx = \sum_{n=1}^{\infty} \sigma_n \langle x, u_n\rangle u_n.$$

Then T is an injective, compact, positive self adjoint operator on H, and for $s > 0$, we have

$$T^s x = \sum_{n=1}^{\infty} \sigma_n^s \langle x, u_n\rangle u_n, \quad x \in H.$$

Thus,

$$\|x\|_s^2 := \sum_{n=1}^{\infty} \frac{|\langle x, u_n\rangle|^2}{\sigma_n^{2s}} = \|T^{-s}x\|^2, \quad x \in R(T^s).$$

Note that

$$R(T^s) := \{x \in H : \sum_{n=1}^{\infty} \frac{|\langle x, u_n\rangle|^2}{\sigma_n^{2s}} \text{ converges}\}.$$

Thus the Hilbert scale $\{H_s : s \in \mathbb{R}\}$ is generated by $L := T^{-1}$. Observe that

$$\langle T^{-1}x, x\rangle = \sum_{n=1}^{\infty} \frac{|\langle x, u_n\rangle|^2}{\sigma_n} \geq \frac{1}{\max_n \sigma_n}\|x\|^2$$

so that T^{-1} is strictly coercive, positive self adjoint operator on $D(T^{-1}) := R(T)$. □

Theorem 2.3 *Let $\{H_s : s \in \mathbb{R}\}$ be as in Example 2.2 and $u_n^{(s)} := \sigma_n^s u_n$ for $s \in \mathbb{R}$ and $n \in \mathbb{N}$.*

(i) *$\{u_n^{(s)} : n \in \mathbb{N}\}$ is an orthonormal basis of H_s.*
(ii) *For $s < t$, the identity map $\mathcal{I}_{s,t} : H_t \to H_s$ is a compact embedding.*

Proof For $x \in H_s$, we have

$$\langle x, u_j^{(s)}\rangle_s = \sum_{n=1}^{\infty} \frac{\langle x, u_n\rangle\langle u_n, u_j^{(s)}\rangle}{\sigma_n^{2s}} = \sum_{n=1}^{\infty} \sigma_i^s \frac{\langle x, u_n\rangle\langle u_n, u_j\rangle}{\sigma_n^{2s}} = \frac{\langle x, u_j\rangle}{\sigma_j^s}. \tag{2.1}$$

Hence,

$$\langle u_i^{(s)}, u_j^{(s)}\rangle_s = \frac{\langle u_i^{(s)}, u_j\rangle}{\sigma_j^s} = \frac{\sigma_i^{(s)}\langle u_i, u_j\rangle}{\sigma_j^s} = \delta_{ij}.$$

Further, for every $x \in H_s$, by (2.1),

$$x = \sum_{n=1}^{\infty}\langle x, u_n\rangle u_n = \sum_{n=1}^{\infty} \sigma_n^s\langle x, u_n^{(s)}\rangle_s\, u_n = \sum_{n=1}^{\infty}\langle x, u_n^{(s)}\rangle_s\, u_n^{(s)}.$$

Hence, $\{u_n^{(s)} : n \in \mathbb{N}\}$ is an orthonormal basis of H_s.

Also, for $x \in H_t$, by (1),

$$x = \sum_{n=1}^{\infty}\langle x, u_n\rangle u_n = \sum_{n=1}^{\infty} \sigma_n^t\langle x, u_n^{(t)}\rangle_t\, u_n = \sum_{n=1}^{\infty} \sigma_n^{t-s}\langle x, u_n^{(t)}\rangle_t\, u_n^{(s)}$$

Thus,

$$\mathcal{I}_{s,t}x = \sum_{n=1}^{\infty} \sigma_n^{t-s}\langle x, u_n^{(t)}\rangle u_n^{(s)}, \quad x \in H_t.$$

Since $\sigma_n^{t-s} \to 0$, the inclusion map is a compact operator. □

Example 2.4 For $s \geq 0$, recall that the Sobolev space

$$H^s(\mathbb{R}^k) := \{f \in L^2(\mathbb{R}^k) : \int_{\mathbb{R}^k} (1+|\xi|^2)^s |\hat{f}(\xi)|^2 d\xi < \infty\}$$

is a Hilbert space with inner product

$$\langle f, g\rangle_s := \int_{\mathbb{R}^k} (1+|\xi|^2)^s \hat{f}(\xi)\overline{\hat{g}(\xi)} d\xi$$

and the corresponding norm

$$\|f\|_s := \left[\int_{\mathbb{R}^k} (1+|\xi|^2)^s |\hat{f}(\xi)|^2 d\xi\right]^{1/2}.$$

For $s < 0$, $H^s(\mathbb{R}^k)$ is defined via Gelfand triple. It can be shown that for $s < t$, the inclusion $H_t \subseteq H_s$ is continuous, and hence $\{H^s(\mathbb{R}^k) : s \in \mathbb{R}\}$ is a Hilbert scale. □

3 Interpolation Inequality in Hilbert Scales

Let $\{H_s\}_{s\in\mathbb{R}}$ be a Hilbert scale. We know that for $r < s < t$, $H_t \subseteq H_s \subset H_r$ and there exist constants $c^{(1)}_{r,s}$ and $c^{(2)}_{s,t}$ such that

$$c^{(2)}_{s,t}\|x\|_t \leq \|x\|_s \leq c^{(1)}_{r,s}\|x\|_r \quad \forall x \in H_t.$$

In most of the standard Hilbert scales, we have another inequality

$$\|x\|_s \leq \|x\|_r^{1-\lambda}\|x\|_t^t, \quad \forall x \in H_t,$$

whenever $r < s < t$, where $\lambda := \frac{s-r}{t-r}$ so that $s = (1-\lambda)r + \lambda t$. This inequality is called the **interpolation inequality** on $\{H_s\}_{s\in\mathbb{R}}$.

Theorem 3.1 *Let $\{H_s\}_{s\in\mathbb{R}}$ be the Hilbert scale as in Example 2.2. For $r < s < t$ and the interpolation inequality*

$$\|x\|_s \leq \|x\|_r^{1-\lambda}\|x\|_t^\lambda, \quad \lambda := \frac{t-s}{t-r}$$

holds for all $x \in H_t$.

Proof Let $r < s < t$ and $u \in H_t$. Since $s = (1-\lambda)r + \lambda t$ with $\lambda := (t-s)/(t-r)$ we write

$$\sum_{n=1}^{\infty} \frac{|\langle x, u_n \rangle|^2}{\sigma_n^{2s}} = \sum_{n=1}^{\infty} \frac{|\langle x, u_n \rangle|^2}{\sigma_n^{2[(1-\lambda)r+\lambda t]}} = \sum_{n=1}^{\infty} \Big[\frac{|\langle x, u_n \rangle|^{2(1-\lambda)}}{\sigma_n^{2(1-\lambda)r}}\Big]\Big[\frac{|\langle u, u_n \rangle|^{2\lambda}}{\sigma_n^{2\lambda t}}\Big].$$

Applying Hölder's inequality by taking $p = 1/(1-\lambda)$ and $q = 1/\lambda$, we have

$$\begin{aligned}\sum_{n=1}^{\infty} \frac{|\langle x, u_n \rangle|^2}{\sigma_n^{2s}} &= \sum_{n=1}^{\infty} \Big[\frac{|\langle x, u_n \rangle|^{2(1-\lambda)}}{\sigma_n^{2(1-\lambda)r}}\Big]\Big[\frac{|\langle u, u_n \rangle|^{2\lambda}}{\sigma_n^{2\lambda t}}\Big] \\ &\le \left\{\sum_{n=1}^{\infty} \frac{|\langle x, u_n \rangle|^2}{\sigma_n^{2r}}\right\}^{1-\lambda} \left\{\sum_{n=1}^{\infty} \frac{|\langle x, u_n \rangle|^2}{\sigma_n^{2t}}\right\}^{\lambda} \\ &= \left\{\|x\|_r^2\right\}^{1-\lambda} \left\{\|x\|_t^2\right\}^{\lambda}.\end{aligned}$$

Thus, $\|x\|_s \le \|x\|_r^{1-\lambda}\|x\|_t^{\lambda}$. □

Theorem 3.2 *Consider the Hilbert scale $\{H^s(R^k)\}_{s\in\mathbb{R}}$ as in Example 2.4. For $r < s < t$, the interpolation inequality*

$$\|f\|_s \le \|f\|_r^{1-\lambda}\|f\|_t^{\lambda}$$

holds for all $f \in H^s(\mathbb{R}^k)$, where λ is such that $s = (1-\lambda)r + \lambda t$, i.e., $\lambda := \frac{t-s}{t-r}$.

Proof Since $s = (1-\lambda)r + \lambda t$, Hölder's inequality gives

$$\begin{aligned}\|f\|_s^2 &= \int_{\mathbb{R}^k} (1+|\xi|^2)^s |\hat{f}(\xi)|^2 d\xi \\ &= \int_{\mathbb{R}^k} [(1+|\xi|^2)^r |\hat{f}(\xi)|^2]^{(1-\lambda)} [(1+|\xi|^2)^t |\hat{f}(\xi)|^2]^{\lambda} d\xi \\ &\le \left[\int_{\mathbb{R}^k} (1+|\xi|^2)^r |\hat{f}(\xi)|^2 d\xi\right]^{1-\lambda} \left[\int_{\mathbb{R}^k} (1+|\xi|^2)^t |\hat{f}(\xi)|^2\right]^{\lambda} d\xi \\ &= \|f\|_r^{2(1-\lambda)}\|f\|_t^{2\lambda}.\end{aligned}$$

Thus, $\|f\|_s \le \|f\|_r^{1-\lambda}\|f\|_t^{\lambda}$. □

Theorem 3.3 *Let $\{H_s\}_{s\in\mathbb{R}}$ be the Hilbert scale as in Example 2.1. For $r < s < t$ and the interpolation inequality*

$$\|x\|_s \le \|x\|_r^{1-\lambda}\|x\|_t^{\lambda}$$

holds for all $x \in H_t$, where λ is such that $s = (1-\lambda)r + \lambda t$, i.e., $\lambda := \frac{t-s}{t-r}$.

Proof In this case, we have

$$\|x\|_s^2 = \langle L^s x, x\rangle = \int_0^\infty \lambda^s d\langle E_\lambda x, x\rangle = \int_0^\infty \lambda^{r(1-\lambda)}\lambda^{t\lambda} d\langle E_\lambda x, x\rangle,$$

Hence, by Hölder's inequality,

$$\|x\|_s^2 \leq \Big(\int_0^\infty \lambda^r d\langle E_\lambda x, x\rangle\Big)^{1-\lambda}\Big(\int_0^\infty \lambda^t d\langle E_\lambda x, x\rangle\Big)^{\lambda} = \|x\|_r^{2(1-\lambda)}\|x\|_t^{2\lambda}.$$

Thus $\|x\|_s \leq \|x\|_r^{1-\lambda}\|x\|_t^{\lambda}$. □

4 Ill-Posed Operator Equations

Let X and Y be Banach spaces. For a given $y \in Y$, consider the problem of finding a solution x of the operator equation

$$F(x) = y, \tag{4.1}$$

where F is a function defined on a subset $D(F)$ of X taking values in Y. According to Hadamard [5], the above problem is said to be **well-posed** if

(1) for every $y \in Y$ there is a solution x,
(2) the solution x is unique, and
(3) the solution depends continuously on the data (y, F), in the sense that if $(\tilde{y}, \tilde{F})$ is a perturbed data which is close to (y, F) in some sense, then the corresponding solution $\tilde{x}$ is close to x.

If it is not a well-posed problem, it is called an **ill-posed** problem. Operator theoretic formulation of many of the inverse problems that appear in science and engineering are ill-posed. Here are two typical examples of ill-posed problems:

Example 4.1 (Compact operator equation) Let X and Y be Banach spaces and $T : X \to Y$ be a compact operator. If $R(T)$ is not closed, then T cannot have a continuous inverse. Hence, the problem of solving the equation

$$Tx = y$$

is an ill-posed problem.

In the setting of Hilbert spaces, the ill-posedness of a compact operator equation can be illustrated with the help of *singular value decomposition*. Suppose X and Y are Hilbert spaces and $K : X \to Y$ is a compact operator of infinite rank. Then it can be represented as

$$Kx = \sum_{n=1}^{\infty} \sigma_n \langle x, u_n \rangle v_n, \quad x \in X,$$

where $\{u_n : n \in \mathbb{N}\}$ is an orthonormal basis of $N(K)^\perp$, $\{v_n : n \in \mathbb{N}\} \subseteq R(K)$ is an orthonormal basis of $N(K^*)^\perp = \overline{R(K)}$, and (σ_n) is a sequence of positive scalars such that $\sigma_n \to 0$ as $n \to \infty$. The above representation of K is called its **singular value representation**. Let $x \in X$ and $y = Kx$. For $k \in \mathbb{N}$, let

$$y_k = y + \sqrt{\sigma_k} v_k, \quad x_k = x + \frac{1}{\sqrt{\sigma_k}} u_k.$$

Then we have

$$Kx_k = y_k \quad \forall k \in \mathbb{N}.$$

Note that, as $k \to \infty$,

$$\|y - y_k\| = \sqrt{\sigma_k} \to 0 \quad \text{but} \quad \|x - x_k\| = \frac{1}{\sqrt{\sigma_k}} \to \infty.$$

As a prototype of a compact operator equation, one may consider the Fredholm integral equation of the first kind,

$$\int_\Omega k(s,t)x(t)\,dt = y(s), \quad x \in X,\ s \in \Omega,$$

where $k(\cdot,\cdot)$ is a nondegenerate kernel in $L^2(\Omega \times \Omega)$, and Ω is a measurable subset of $\mathbb{R}^k$. Then the operator $T : L^2(\Omega) \to L^2(\Omega)$ defined by

$$(Tx)(s) = \int_\Omega k(s,t)x(t)\,dt, \quad x \in X,\ s \in \Omega,$$

is a compact operator with nonclosed range. Thus, the problem of solving such integral equations is ill-posed.

It may be remarked that Fredholm integral equations of the first kind appears in many inverse problems of practical importance [3]; for example, problems in Computerized tomography, Geophysical prospecting, and Image reconstruction problems. □

Example 4.2 (Parameter identification problem in PDE) Let Ω be a bounded domain in $\mathbb{R}^n$ for some $n \in \mathbb{N}$ and $q(\cdot) \in L^\infty(\Omega)$ be such that $q(\cdot) \geq c_0$ a.e. for some $c_0 > 0$. Then for every $f \in L^2(\Omega)$, there exists a unique $u \in H_0^1(\Omega)$ such that

$$-\nabla.(q(x)\nabla u) = f$$

is satisfied in the weak sense. Now, for a fixed $f \in L^2(\Omega)$, the map

$$F : q \mapsto u$$

defined on the subset

$$D(F) := \{q \in L^\infty(\Omega) : q \geq c_0 \text{ a.e.}\}$$

is a nonlinear operator which does not have a continuous inverse. Thus, the problem of solving the nonlinear equation

$$F(q) = u$$

is an ill-posed equation.

Such ill-posed nonlinear problems appear in many applications, e.g., diffraction tomography, impedance tomography, oil reservoir simulation, and aquifer calibration (see, e.g., [1, 2, 17]).

Let us illustrate the nonlinearity and ill-posedness of the above problem using its one-dimensional formulation:

$$\frac{d}{dt}\left[q(t)\frac{du}{dt}\right] = f(t), \quad 0 < t < 1,$$

where $f \in L^2(0, 1)$. Note that

$$u(t) = \int_0^t \left[\frac{1}{q(\tau)}\int_0^\tau f(s)ds\right] d\tau.$$

Thus, the problem is same as that of solving the equation

$$F(q)(t) := \int_0^t \left[\frac{1}{q(\tau)}\int_0^\tau f(s)ds\right] d\tau = u(t),$$

where $q \in L^\infty(0, 1)$ with $q \geq q_0$ a.e. for some $q_0 > 0$. Clearly, this equation is nonlinear. Note also that

$$q(t) = \frac{1}{u'(t)}\int_0^t f(s)ds.$$

Suppose $u(t)$ is perturbed to $\tilde{u}(t)$, say

$$\tilde{u}(t) = u(t) + \varepsilon(t).$$

Suppose $\tilde{q}(t)$ is the corresponding solution. Then we have

$$\begin{aligned} q(t) - \tilde{q}(t) &= \left[\frac{1}{u'(t)} - \frac{1}{u'(t) + \varepsilon'(t)}\right] \int_0^t f(s)ds \\ &= \frac{1}{u'(t)} \left[\frac{\varepsilon'(t)}{u'(t) + \varepsilon'(t)}\right] \int_0^t f(s)ds. \end{aligned}$$

Hence,

$$\varepsilon'(t) \approx \infty \quad \Longrightarrow \quad |q(t) - \tilde{q}(t)| \approx |q(t)|.$$

There can be perturbations $\varepsilon(t)$ such that

$$\varepsilon(t) \approx 0 \quad \text{but} \quad \varepsilon'(t) \approx \infty.$$

For example, for large n,

$$\varepsilon_n(t) = (1/n)\sin(n^2 x) \approx 0 \quad \text{but} \quad \varepsilon_n'(t) = n\cos(n^2 x) \approx \infty.$$

Thus, the problem is ill-posed. □

5 Regularization

For an ill-posed problem (4.1) with $(\tilde{y}, \tilde{F})$ in place of (y, F), one looks for a family $\{\tilde{x}_\alpha\}_{\alpha>0}$ of approximate solutions such that each $\tilde{x}_\alpha$ is a solution of a well-posed problem and $\alpha := \alpha(\tilde{y}, \tilde{F})$ is chosen in such a way that

$$\tilde{x}_\alpha \to x \quad \text{as} \quad (\tilde{y}, \tilde{F}) \to (y, F).$$

The procedure of finding such a stable approximate solution is called a **regularization method**.

5.1 Tikhonov Regularization

When X and Y are Hilbert spaces and $\tilde{F} = F$, *Tikhonov regularization* is one such regularization methods which is widely used in applications. In Tikhonov regularization, one looks of an approximate solution $\tilde{x}_\alpha$ which is a minimizer of the *Tikhonov functional*

$$J_\alpha : x \mapsto \|F(x) - \tilde{y}\|^2 + \alpha\|x - x^*\|^2.$$

Here, x^* is an initial guess of the solution which can be taken as 0 in the linear case. For the case of a linear ill-posed operator equation,

$$Tx = y$$

with $x^* = 0$, the above regularized problem, that is, the problem of finding the minimizer of

$$x \mapsto \|Tx - \tilde{y}\|^2 + \alpha\|x\|^2$$

has a unique solution and it satisfies the well-posed operator equation

$$(T^*T + \alpha I)\tilde{x}_\alpha = T^*\tilde{y}.$$

It is known (see e.g., Nair [15]) that if $\|\tilde{y} - y\| \leq \delta$ for some $\delta > 0$, the best possible error estimate is

$$\|\hat{x} - \tilde{x}_\alpha\| = O(\delta^{2/3})$$

which is order optimal for the source set

$$\{x \in X : x = (T^*T)u : \|u\| \leq \rho\}$$

and it is attained by an a priori choice of α, namely, $\alpha \sim \delta^{2/3}$ or by the a posteriori choice of Arcangeli's method (see Nair [11]),

$$\|T\tilde{x}_\alpha - \tilde{y}\| = \frac{\delta}{\sqrt{\alpha}}.$$

5.2 Improvement Using Hilbert Scales

Now, the question is whether we can modify the above procedure, probably by requiring more regularity for the regularized solution and the sought for unknown solution, to yield better order in the error estimate. That is exactly what Natterer [16] suggested using Hilbert scales. Natterer's idea was to look for a modification of the Tikhonov regularization which yield an approximation of the LRN-solution which minimizes the function

$$x \mapsto \|x\|_s,$$

where $\|\cdot\|_s$ for $s > 0$ is the norm on the Hilbert space H_s corresponding to a Hilbert scale $\{H_s : s \in \mathbb{R}\}$ for which the interpolation inequality

$$\|u\|_s \leq \|u\|_r^{1-\lambda}\|u\|_t^{\lambda}, \quad \lambda := \frac{t-s}{t-r}$$

holds for $r \le s \le t$. This purpose was served by considering the minimizer $\tilde{x}_{\alpha,s}$ of

$$x \mapsto \|Tx - \tilde{y}\|^2 + \alpha\|x\|_s^2.$$

Natterer showed that if T satisfies

$$\|Tx\| \ge c\|x\|_{-a} \quad \forall x \in X$$

for some $a > 0$ and $c > 0$, and if $\hat{x} \in H_t$ where $0 \le t \le 2s+a$ and the regularization parameter α is chosen such that $\alpha \sim \delta^{\frac{2(a+s)}{a+t}}$, then

$$\|\tilde{x}_{\alpha,s} - \hat{x}\| = O(\delta^{\frac{t}{t+a}}). \tag{5.1}$$

Thus, higher smoothness requirement on $\hat{x}$ and with higher level of regularization gives higher order of convergence.

5.3 Further Improvements Under Stronger Source Conditions

For obtaining further improvements on the error estimates, another modification considered extensively in the literature was to look for an approximation of the LRN-solution which minimizes the function

$$x \mapsto \|Lx\|,$$

where $L : D(L) \subseteq X \to X$ is a closed densely defined operator. It is known (cf. Nair et al. [12]) that such an LRN-solution exists whenever

$$y \in R(T_{|D(L)}) + R(T)^{\perp}.$$

Accordingly, the associated modification in the regularization method is done by looking for the unique minimizer of the function

$$x \mapsto \|Tx - \tilde{y}\|^2 + \alpha\|Lx\|^2, \quad x \in D(L),$$

Throughout this paper, we shall assume that $R(T)$ is not closed and (T, L) satisfies the following condition:

Completion condition: The operators T and L satisfy the condition

$$\|Tx\|^2 + \|Lx\|^2 \ge \gamma\|x\|^2 \quad \forall x \in D(L), \tag{5.2}$$

for some positive constant γ.

Regarding the existence, uniqueness and convergence of the regularized solutions, we have the following theorem (cf. Locker and Prenter [6], Nair et al. [12]).

Theorem 5.1 *Suppose the operators T and L satisfy the condition (5.2). Then for every $y \in Y$ and $\alpha > 0$, the function*

$$x \mapsto \|Tx - \tilde{y}\|^2 + \alpha\|Lx\|^2, \quad x \in D(L),$$

has a unique minimizer and it is the solution $x_\alpha(y)$ of the well-posed equation

$$(T^*T + \alpha L^*L)x = T^*y.$$

Further, if $y \in R(T_{|D(L)}) + R(T)^\perp$, then

$$x_\alpha(y) \to \hat{x} \quad as \quad \alpha \to 0,$$

where $\hat{x}$ is the unique LRN-solution which minimizes the map $x \mapsto \|Lx\|$.

Note that for $L = I$, we obtain the ordinary Tikhonov regularization, and (5.2) is satsifed, if for example L is bounded below, which is the case for many of the differential operators that appear in applications.

When we have the perturbed data y^δ with

$$\|y - \tilde{y}\| \leq \delta,$$

the regularized solution is the minimizer of

$$x \mapsto \|Tx - \tilde{y}^\delta\|^2 + \alpha\|Lx\|^2, \quad x \in D(L);$$

equivalently, the solution x_α^δ of the well-posed equation

$$(T^*T + \alpha L^*L)x_\alpha^\delta = T^*y^\delta.$$

In this case, it is required to choose the regularization parameter $\alpha := \alpha(\delta, y^\delta)$ appropriately so that

$$x_\alpha^\delta \to \hat{x} \text{ as } \alpha \to 0.$$

In order to obtain error estimates, it is necessary to impose some smoothness assumptions on $\hat{x}$, by requiring it to belong to certain *source set*. This aspect has been considered extensively in the literature in recent years by assuming that the operator L is associated with a Hilbert scale $\{X_s\}_{s\in\mathbb{R}}$ in an appropriate manner (see, e.g., [7, 13, 16]). One such relation is as in the following.

Hilbert scale conditions:

(i) There exists $a > 0, c > 0$ such that

$$\|Tx\| \geq c\|x\|_{-a} \quad \forall x \in X. \tag{5.3}$$

(ii) There exists $b \geq 0, d > 0$ such that $D(L) \subseteq X_b$ and

$$\|Lx\| \geq d\|x\|_b \quad \forall x \in D(L). \tag{5.4}$$

Theorem 5.2 (Nair [10]) *If the Hilbert scale conditions (5.3) and (5.4) are satisfied and if $\hat{x}$ belongs to the source set*

$$M_\rho = \{x \in D(L) : \|Lx\| \leq \rho\} \tag{5.5}$$

for some $\rho > 0$, α is chosen according to the Morozov discrepancy principle

$$c_1\delta \leq \|Tx_\alpha^\delta - y^\delta\| \leq c_0\delta \tag{5.6}$$

with $c_0, c_1 \geq 1$, then

$$\|\hat{x} - \tilde{x}_\alpha\| \leq 2\left(\frac{\rho}{d}\right)^{\frac{a}{a+b}}\left(\frac{\delta}{c}\right)^{\frac{b}{a+b}}. \tag{5.7}$$

Remark 5.3 The estimate in (5.7) corresponds to the estimate (5.1) obtained by Natterer for the case $t = s = b$. The discrepancy principle considered in Nair [10] was $\|Tx_\alpha^d - y^\delta\| = \delta$, which can be easily modified to (5.6). ◇

For obtaining further improved estimate, two more source sets are considered in Nair [14], namely,

$$\widetilde{M}_\rho = \{x \in D(L) : \|L^*Lx\| \leq \rho\}, \tag{5.8}$$

$$M_{\rho,\varphi} := \{x \in D(L^*L) : \ L^*Lx = [\varphi(T^*T)]^{1/2}u, \ \|u\| \leq \rho\} \tag{5.9}$$

for some constant $\rho > 0$, and for some *index function*, i.e., a strictly monotonically increasing continuous function $\varphi : [0, \infty) \to [0, \infty)$ such that $\lim_{\lambda \to 0} \varphi(\lambda) = 0$, which is also concave.

Theorem 5.4 (Nair [14]) *Suppose the Hilbert scale conditions (5.3) and (5.4) are satisfied and α is chosen according to the Morozov discrepancy principle (5.6).*

(i) *If $\hat{x}$ belongs to the source set $\widetilde{M}_\rho$ defined in (5.8), then*

$$\|\hat{x} - \tilde{x}_\alpha\| \leq 2\left(\frac{\rho}{d^2}\right)^{\frac{a}{2a+b}}\left(\frac{\delta}{c}\right)^{\frac{2b}{a+2b}}.$$

(ii) *If $\hat{x}$ belongs to the source set $M_{\rho,\varphi}$ defined in (5.9), then*

$$\|\hat{x} - \tilde{x}_\alpha\| = (1 + c_0)\left(\frac{\rho}{d^2}\right)^{\frac{a}{a+2b}} \left(\frac{\delta}{c}\right)^{\frac{2b}{a+b}} \left[\psi_p^{-1}\left(\varepsilon_\delta^2\right)\right]^{\frac{a}{2(a+2b)}}$$

where $p = a/(a + 2b)$, $\varepsilon_\delta := c\left(\frac{d^2\delta}{c\rho}\right)^{\frac{a}{a+2b}}$ *and* $\psi_p(\lambda) := \lambda^{1/p}\varphi^{-1}(\lambda)$.

In a recent paper [8], the author has obtained results by replacing the Hilbert scale conditions (5.3) and (5.4) by a single condition involving T and T, as given below.

θ-condition:
There exist $\eta > 0$ and $0 \le \theta < 1$ such that

$$\eta\|x\| \le \|Tx\|^\theta \|Lx\|^{1-\theta} \quad \forall x \in D(L). \tag{5.10}$$

We may observe that the case of $\theta = 0$ corresponds to L being bounded below so that $R(L)$ is closed and $L^{-1} : R(L) \to X$ is a bounded operator. This case also include the choice $L = I$, the identity operator. Note that the value $\theta = 1$ is excluded, as it would imply that T has a continuous inverse.

It has been shown that the θ-condition implies the completion condition (5.2) with $\gamma = \eta^2$, and it is implied by the Hilbert scales conditions (5.3) and (5.4) with

$$\theta = \frac{b}{a + b}, \qquad \eta = c^{\frac{b}{a+b}} d^{\frac{a}{a+b}}.$$

Among other results, the following theorem has been proved in Nair [8], which unifies results in the setting of general unbounded stabilizing operator as well as for Hilbert scale and Hilbert scale-free settings.

Theorem 5.5 (Nair [8]) *Suppose the θ-condition (5.10) is satisfied and α is chosen according to the Morozov discrepancy principle (5.6).*

(i) *If $\hat{x}$ belongs to the source set M_ρ defined in (5.5), then*

$$\|\hat{x} - x_{\alpha_\delta}^\delta\| \le 2\eta^{-1}\rho^{1-\theta}\delta^\theta.$$

(ii) *If $\hat{x}$ belongs to the source set $\widetilde{M}_\rho$ defined in (5.8), then*

$$\|\hat{x} - x_{\alpha_\delta}^\delta\| \le (1 + c_0)\left(\frac{1}{\eta^2}\right)^{\frac{1}{1+\theta}} \rho^{\frac{1-\theta}{1+\theta}} \delta^{\frac{2\theta}{1+\theta}}.$$

(iii) *If $\hat{x}$ belongs to the source set $M_{\rho,\varphi}$ defined in (5.9) and $\delta^2 \le \gamma_1^2\varphi(1)$ where $\gamma_1 = 4\rho/\gamma$, then*

$$\|\hat{x} - x_{\alpha_\delta}^\delta\| \le (1 + c_0)\left(\frac{1}{\eta^2}\right)^{\frac{1}{1+\theta}} \rho^{\frac{1-\theta}{1+\theta}} \delta^{\frac{2\theta}{1+\theta}} [\psi_p^{-1}(\varepsilon_\delta^2)]^{1/2p},$$

where

$$p := \frac{1+\theta}{1-\theta}, \quad \varepsilon_\delta := \eta^{\frac{2}{1+\theta}} \left(\frac{\delta}{\rho}\right)^{1/p}.$$

Remark 5.6 The results in Theorems 5.2 and 5.4 are recovered from Theorem 5.5 by taking $\theta = \dfrac{b}{a+b}$ and $\eta = c^{\frac{b}{a+b}} d^{\frac{a}{a+b}}$.

In the ordinary Tikhonov regularization, i.e., for the case of $L = I$, equivalently, $\theta = 0$, in part (iii) of the above theorem, we have $p = 1$ and $\varepsilon_\delta = \eta^2 \left(\frac{\delta}{\rho}\right)$. Hence, the estimate reduces to

$$\|\hat{x} - x^\delta_{\alpha_\delta}\| \le (1 + c_0) \left(\frac{1}{\eta^2}\right) \sqrt{\psi_1^{-1}(\eta^2\delta^2/\rho^2)}.$$

Thus, from the above, we recover the error estimate under the general source condition derived in [7]. ◇

References

1. Borcea, L., Berryman, J.G., Papanicolaou, G.C.: High-contrast impedance tomography. Inverse Probl. **12** (1996)
2. Devaney, A.J.: The limited-view problem in diffraction tomography. Inverse Probl. **5**(5) (1989)
3. Engl, H.W., Hanke, M., Neubauer, A.: Regularization of Inverse Problems. Kluwer, Dordrecht (2003)
4. Groetsch, C.W.: The Theory of Tikhonov Regularization for Fredholm Equations of the First Kind. Pitman Pub, Melbourne (1984)
5. Hadamard, J.: Lectures on the Cauchy Problem in Linear partial Differential Equations, Yale University Press (1923)
6. Locker, J., Prenter, P.M.: Regularization with differential operators. Math. Anal. Appl. **74**, 504–529 (1980)
7. Nair, M.T., Schock, M.E., Tautenhahn, U.: Morozov's discrepancy principle under general source conditions, J. Anal. Anw. **22**, 199–214 (2003)
8. Nair, M.T.: A Unified Treatment for Tikhonov Regularization Using a General Stabilizing Operator, Analysis and Applications, (2014) (To appear)
9. Nair, M.T.: Functional Analysis: A First Course. Printice-Hall of India, New Delhi (2002) (Forth Print: 2014)
10. Nair, M.T.: On Morozov's method Tikhonov regularization as an optimal order yielding algorithm. Zeit. Anal. und ihre Anwendungen, 37–46, **18**(1) (1999)
11. Nair, M.T.: A generalization of Arcangeli's method for ill-posed problems leading to optimal rates. Integral Equ. Operator Theory **15**, 1042–1046 (1992)
12. Nair, M.T., Hegland, M., Anderssen, R.S.: The trade-off between regularity and stability in Tikhonov regularization. Math. Comput. **66**, 193–206 (1997)
13. Nair, M.T., Pereverzyev, M.S., Tautenhahn, U.: Regularization in Hilbert scales under general smoothing conditions. Inverse Probl. **21**, 1851–1869 (2003)
14. Nair, M.T.: On improving esitmates for Tikhonov regularization using an unbounded opertor. J. Anal. **14**, 143–157 (2006)
15. Nair, M.T.: Linear Operator Equations: Approximation and Regularization. World Scientific, New York (2009)

16. Natterer, F.: Error bounds for Tikhonov regularization in Hilbert scales. Appl. Anal. **18**, 29–37 (1984)
17. Parker, R.L.: Geophysical Inverse Theory. Princeton University Press, Princeton NJ (1994)

On Three-Space Problems for Certain Classes of C^*-algebras

A.K. Vijayarajan

Abstract It is shown that being a GCR algebra is a three-space property for C^*-algebras using the structure of composition series of ideals present in GCR algebras. A procedure is presented to construct a composition series for a C^*-algebra from the unique composition series for any GCR ideal and the corresponding GCR quotient being a C^*-algebra. We deduce as a consequence that, a GCR algebra is a three-space property. While noting that being a CCR algebra is not a three-space property for C^*-algebras, sufficient additional conditions required on a C^*-algebra for the CCR property to be a three-space property are also presented. Relevant examples are also presented.

Keywords CCR algebras · GCR algebras · Three-space problem · C^*-algebra

Mathematics Subject Classification (2010) Primary 46L05 · Secondary 46L35

1 Introduction

In the Banach space setting where the three-space problem originally appeared investigated properties that a Banach space shares under the assumption that a closed subspace and the corresponding quotient space have the property. If the answer to the problem is affirmative, then the property is called a three-space property for Banach spaces. Though Krein and Smullian essentially proved that reflexivity is a three-space property (without calling it as such) for Banach spaces; see [7], three-space problem

This article is an expanded form of an invited talk delivered on 27 February 2014 during the *International Conference on Semigroups, Algebras and Operator Theory (ICSAOT) held during February 26–28, 2014 at the Cochin University of Science and Technology.*

A.K. Vijayarajan (✉)
Kerala School of Mathematics, Kunnamangalam (PO),
Kozhikode (DT) 673571, Kerala, India
e-mail: vijay@ksom.res.in

P.G. Romeo et al. (eds.), *Semigroups, Algebras and Operator Theory*,
Springer Proceedings in Mathematics & Statistics 142,
DOI 10.1007/978-81-322-2488-4_14

did not formally appear until the middle 1970s when Enflo, Lindenstrauss and Pisier settled in the negative a question of Palais: If $Y \subset X$ and X/Y are Hilbert spaces, is X isomorphic to a Hilbert space; see [4].

Several three-space problems settled in the affirmative and negative exist in Banach space setting. We will cite a few of them here. In [5], it is proved that admitting a locally uniformly reflexive norm is a three-space property, whereas in [8], the problem is settled in the negative for properties, 'to be complemented in a dual space' and 'to be isomorphic to a dual space'. For a detailed discussion of the three-space problem for Banach spaces and related topics, we refer to [2]. In [9] three-space problems are settled in the affirmative and the negative for certain classes of L^1-preduals and also for general Banach spaces with the additional assumption that the subspace in question is an M-ideal.

Now we look at the three-space problem in the context of C^*-algebras where the problem needs to be reformulated appropriately.

As several important properties of C^*-algebras are often completely characterized by the ideal reduction condition that both the ideal and the corresponding quotient should possess the same property as the original C^*-algebra, it is important and interesting to address the three-space problems in the theory of C^*-algebras. But, in the C^*-algebra setting, the natural formulation of the three-space problem is as follows:

Let A be a C^*-algebra and let I be a closed two-sided ideal of A. Is it true that A has property P when both I and the quotient A/I have the property P? As before, if the answer to the question is affirmative, then we call P a three-space property for C^*-algebras.

Three-space problem in the setting of C^*-algebras also has results for important properties. For example, a C^*-algebra A is of type I if and only if both I and the quotient A/I are of type I, or A is nuclear if and only if both I and the quotient A/I are nuclear and so on (see [3]), whereas to be a dual C^*-algebra is not a three-space property (see [6]).

In this work, we study three-space problem for GCR and CCR properties of C^*-algebras. We show that for C^*-algebras, being a GCR algebra is a three-space property. While observing that being a CCR algebra is not a three-space property for C^*-algebras, we investigate conditions so that a C^*-algebra is a CCR algebra if an ideal of it and the corresponding quotient are CCR algebras.

In this introductory section, we establish our notation and terminology for general C^*-algebras and use the next section to discuss GCR and CCR algebras before addressing the three-space problem for them in the last section. We refer to [1] for our terminology and notation on C^*-algebras.

Let A be a C^*-algebra. A *representation* π of A on a Hilbert space H is a $*$-homomorphism of A into the C^*-algebra $L(H)$ of all bounded operators on H. We call π to be *non-degenerate* if the C^*-algebra of operators $\pi(A)$ has trivial null space which is equivalent to saying that the closed linear span of $\{\pi(x)\xi : x \in A, \xi \in H\} = [\pi(A)H] = H$.

Two representations π and σ of A represented on Hilbert spaces H and K, respectively, are said to be *(unitarily)equivalent* if there exists a unitary operator $U : H \mapsto K \ni \sigma(x) = U\pi(x)U^*, \forall x \in A$. A non-zero representation π of A is called *irreducible* if $\pi(A)$ is an irreducible operator algebra. i.e., it commutes with no non-trivial (self-adjoint) projections. Since $\pi(A)$ is a C^*-algebra, irreducibility is equivalent to saying that $\pi(A)$ has no non-trivial closed invariant subspaces.

The ideals we consider here are closed two-sided ideals. In general, a representation of a subalgebra of a C^*-algebra on a Hilbert space H cannot be extended to to a representation of the whole algebra A on H. But, if the subalgebra happens to be an ideal, then extension is possible. In fact, we have, if J is a ideal of A and π a non-degenerate representation of J on a Hilbert space H, then for each $x \in A$, there is a unique bounded linear operator $\tilde{\pi}(x)$ on H satisfying $\tilde{\pi}(x)\pi(y) = \pi(xy), \forall y \in J$ (see [1]). The restriction to non-degenerate representation is not a serious one as, in the general case, we can pass from H to the subspace $[\pi(J)H]$, where we can still get a unique extension $\tilde{\pi}$ of π to $A \ni \tilde{\pi}(A)$ acts on $[\pi(J)H]$.

An operator T $\in L(H)$ is said to be compact if T(U) has compact closure, where U is the unit ball in H. The set $C(H)$ of all compact operators on H is an ideal in $L(H)$ and is therefore a C^*-algebra in its own right. Recall that $C(H)$ has no non-trivial proper irreducible subalgebras and as a result it has no non-trivial ideals. Also, any irreducible representation of $C(H)$ is equivalent to the identity representation of $C(H)$.

2 C^*-algebras of Type CCR and GCR

In this section we introduce the important classes of CCR and GCR algebras. We also present the very important notion of composition series of ideals in a C^*-algebra which are very relevant for our classes. Examples of these algebras are also given to explain how they are related.

Definition 2.1 A CCR algebra is a C^*-algebra A such that for every irreducible representation π of A, $\pi(A)$ consists of compact operators.

Clearly, every C^*-algebra of compact operators is CCR. Also, every commutative C^*-algebra is CCR since each of its irreducible representation is one dimensional.

Let A be a general C^*-algebra and π be an irreducible representation of A on a Hilbert space H. we define : $C_\pi = \{x \in A : \pi(x) \in C(H)\}$ and CCR(A) $= \bigcap_\pi C_\pi$, where the intersection ranges over all irreducible representations π of A. Thus CCR(A) is the maximal CCR ideal in A.

Definition 2.2 A C^*-algebra A is said to be GCR if CCR(A/I) $\neq 0$ for every ideal I of A.

Any irreducible representation of the quotient A/I of a C^*-algebra A may be composed with the quotient map of A into A/I. It follows that every quotient of a CCR algebra is CCR and hence every CCR algebra is GCR. In fact $\mathrm{CCR}(A/I) = A/I$ if A is CCR.

The definition of a GCR algebra we have given above is not a convenient one to work with as one has to find all ideals I in a given algebra A before examining the quotients A/I. The following result gives a rather simple characterization for GCR algebras similar to the definition of CCR algebras, but it is very hard to prove.

Theorem A (see [3]) *A C^*-algebra A is GCR if and only if for every irreducible representation π of A, $\pi(A)$ contains a non-zero compact operator and hence all of them.*

We will give another structural condition equivalent to the GCR property, which can be easily proved using the definition of GCR algebras only. Before stating the theorem we need the following definition.

Definition 2.3 A composition series in a C^*-algebra A is a family of ideals $\{J_\alpha : 0 \leq \alpha \leq \alpha_0\}$ indexed by the ordinals α, $0 \leq \alpha \leq \alpha_0$, having the following properties:
(i) For all $\alpha \leq \alpha_0$, J_α is contained properly in $J_{\alpha+1}$;
(ii) $J_0 = 0$ and $J_{\alpha_0} = A$;
(iii) If β is a limit ordinal then J_β is the norm closure of $\bigcup_{\alpha<\beta} J_\alpha$.

Theorem B (see [1]) *Every GCR algebra A has exactly one composition series $\{J_\alpha : 0 \leq \alpha \leq \alpha_0\}$ with the property that $J_{\alpha+1}/J_\alpha$ is the largest CCR ideal in A/J_α for every $\alpha, 0 \leq \alpha \leq \alpha_0$. Conversely, if A admits a composition series $\{J_\alpha : 0 \leq \alpha \leq \alpha_0\}$ such that each quotient $J_{\alpha+1}/J_\alpha$ is CCR, then A is GCR.*

The CCR, GCR and C^*scenario are illustrated with a couple of examples below.

Example 2.4 An example of a C^*-algebra which is GCR but not CCR:

Let H be an infinite dimensional Hilbert space and T an irreducible non-compact operator on H with its imaginary part compact. Then, the C^*-algebra $C^*(T)$ generated by T and the identity operator 1 is clearly not CCR. However, since imaginary part of T is compact, $C^*(H) \subset C^*(T)$. Further, $C^*(T)/C(H)$ is commutative and hence CCR. Thus, $\{0, C(H), C^*(T)\}$ is a composition series for $C^*(T)$ with CCR quotients. Hence $C^*(T)$ is GCR.

By taking $A = C(H) + \mathbb{C}1$ and $I = C(H)$ the above situation is explained in a simpler way.

Example 2.5 An example of a C^*-algebra which is not GCR:

Let H be an infinite dimensional Hilbert space. Consider the C^*-algebra $L(H)$. Note that $C(H)$ is the only nontrivial ideal in $L(H)$ and so $K(H) = L(H)/C(H)$ is simple. So $\{0, C(H), L(H)\}$ is the only composition series for $L(H)$. Since H is infinite dimensional and $K(H)$ is simple, we conclude that for any irreducible representation π of $K(H)$, $\pi(e)$ cannot be compact, where e is the identity element of $K(H)$. So $K(H)$ is not CCR. Hence $L(H)$ is not GCR.

3 CCR, GCR Algebras and Three-Space Property for C^*-algebras

Here we ask whether being a CCR algebra and being a GCR algebra are three-space properties for C^*-algebras.

The CCR case is considered first where we have a negative answer to our question by citing a counterexample and then look for additional conditions to get an affirmative answer.

The Example 2.4 can be used to show that being a CCR algebra is not a three-space property for C^*-algebras. Recall that the ideal $C(H)$ and the quotient $C^*(T)/C(H)$ are CCR algebras, whereas $C^*(T)$ in not a CCR algebra. We look for some additional conditions on A which will make A a CCR algebra. We explore the topological properties of the spectrum of A for achieving this.

We denote by $\hat{A}$ the spectrum of A, that is, the equivalence classes of non-zero irreducible representations of A equipped with the Jacobson topology. Since irreducible representations of an ideal I of A extends uniquely to an irreducible representation of A, we may identify $\hat{I}$ as a subset of $\hat{A}$. In fact $I \to \hat{I}$ is a one-to-one mapping from the collection of all closed ideals of A onto the collection of all open sets in $\hat{A}$. Also, $\widehat{A/I}$ may be identified with the closed set $\hat{A}/\hat{I}$.

It is known that a GCR algebra A is CCR if and only if its dual $\hat{A}$ is a T_1 topological space. Also, it can be shown that if the open central projection corresponding to I, satisfying $I = \mathrm{p}A^{**} \cap I$ is a multiplier for A, then $\widehat{A/I}$ is open in $\hat{A}$ (see [6], Lemma 2.1). Thus we have the following result:

Proposition 3.1 *Let A be a C^*-algebra and I be an ideal of A such that I and A/I are CCR and the open central projection corresponding to I is a multiplier of A. Then A is also CCR.*

Proof Since I and A/I are CCR, both $\hat{I}$ and $\widehat{A/I}$ are T_1. Also, since the open central projection corresponding to I is a multiplier for A, $\widehat{A/I}$ is open. Hence $\hat{A} = \hat{I} \cup \widehat{A/I}$ is T_1 and thus A is CCR. □

Now we consider the GCR case and we have the following theorem.

Theorem 3.2 *A C^*-algebra A is GCR if and only if there exist an ideal I of A such that both I and the quotient A/I are GCR. Thus the property of being a GCR algebra is a three-space property.*

Proof Let A be a GCR algebra. Then by definition CCR$(A/I) \neq 0$ for every ideal I of A. In particular CCR$(A) \neq 0$. Let I = CCR(A), which is clearly GCR. By Theorem B, A has a unique composition series $\{J_\alpha{:}0 \leq \alpha \leq \alpha_0\}$ such that $J_{\alpha+1}/J_\alpha$ = CCR(A/J_α). Then $\{J_\alpha/I : 0 \leq \alpha \leq \alpha_0\}$ is the composition series for A/I with $J_{\alpha+1}/I/J_\alpha/I$ = CCR$(A/I/J_\alpha/I)$. Hence A/I is also GCR.

Conversely, assume that A has an ideal I such that both I and A/I are GCR. Let $\{I_\alpha : 0 \leq \alpha \leq \alpha_0\}$ and $\{J_\alpha/I : 0 \leq \alpha \leq \alpha_0\}$ be composition series for I and A/I, respectively, with the quotients $I_{\alpha+1}/I_\alpha$ and $J_{\alpha+1}/I/J_\alpha/I$ CCR (here J_α are ideals in A). If we are able to cook up a composition series for A using these two composition series, we will be done.

For this, first we will show that I_α are ideals in A also.
Let x $\in A$ and y $\in I_\alpha$ then xy $\in I$.
Let $e_1, e_2, e_3, \ldots$ be the local approximate identity for y in I_α so that lim $||ye_n - y||$ = 0.
Now, $||xye_n - xy|| \leq ||x||||ye_n - y||$ and
therefore lim $||xye_n - xy|| = 0$.
Since $xye_n \in I_\alpha$, we see that $xy \in I_\alpha$.
Hence I_α are ideals in A also. Thus $\{ I_\alpha : 0 \leq \alpha \leq \alpha_0\} \cup \{J_\alpha : 0 \leq \alpha \leq \alpha_0\}$ is a composition series for A with the respective quotients $I_{\alpha+1}/I_\alpha$ and $J_{\alpha+1}/J_\alpha$ CCR. □

Remark 3.3 If A is a C^*-algebra and I is a GCR ideal of A such that A/I is also GCR, then by Theorem A we can easily conclude that A is also GCR. Here in the above theorem, we make use of the concept of composition series for GCR algebras which captures the ideal structure of the algebra completely to deduce the same result.

Remark 3.4 In Theorem 3.2, if we start with the unique composition series for I and A/I, then the resulting composition series for A will be its unique composition series if and only is CCR(A/I_α) = CCR(I/I_α), for all $\alpha < \alpha_0$. Note that CCR(A) = $CCR(I)$ need not imply that CCR$(A/I_\alpha) = CCR(I/I_\alpha)$. For example, let T be the operator in Example 2.4. Let $A = C^*(T)$ and let I be the ideal generated by T^2 in A. Then $C(H) \subset I \subset A$. So CCR$(A/C(H)) = A/C(H)$ whereas CCR$(I/C(H)) = I/C(H)$.

In the following theorem we employ the tool of composition series to easily deduce that ideals and associated quotients of a GCR algebra inherit the GCR property.

Theorem 3.5 *If A is a GCR algebra and I is an ideal in A, then I and A/I are GCR algebras.*

Proof Suppose that A is a GCR algebra and I is an ideal in A. Let $\{J_\alpha : 0 \leq \alpha \leq \alpha_0\}$ be a composition series for A such that the quotients $J_{\alpha+1}/J_\alpha = CCR(A/J_\alpha)$. Let $I_\alpha = J_\alpha \cap I$, and α_1 be that first ordinal $\ni I_\alpha = I$. we claim that $\{I_\alpha : 0 \leq \alpha \leq \alpha_1\}$ is a composition series for I. For let $\beta < \alpha_1$ be a limit ordinal. Then the norm closure of $\bigcup_{\alpha<\beta} J_\alpha = J_\beta$. Clearly, $\bigcup_{\alpha<\beta} I_\alpha \subseteq I_\beta$ and hence the norm closure of $\bigcup_{\alpha<\beta} I_\alpha \subseteq I_\beta$.

To get the reverse inclusion, let x $\in I_\beta$ be a self-adjoint element and $\epsilon > 0$. Then there exist a self-adjoint y$\in J_\alpha$ for some $\alpha < \alpha_0$ such that $||x - y|| < \epsilon/2$ (if y is not self-adjoint replace y by (y + y*)/2). Let f be a continuous real valued function with f(t) = 0 for $|t| < \epsilon/2$, f(t) = t for $|t| \geq \epsilon$ and linear in between. Then f(x)$\in I$, f(y)$\in J_\alpha$ and f(x – y) = 0 so that f(x)$\in I_\alpha$. Also $||f(x) - x|| = ||f(y) - x|| = 3||y||/\epsilon||(\epsilon y/3||y|| - \epsilon x/3||y||)|| = ||x - y|| < \epsilon$. Therefore, x is in the norm closure of $\bigcup_{\alpha<\beta} I_\alpha$. Hence $\{I_\alpha : 0 \leq \alpha \leq \alpha_0\}$ is a composition series for I. Also, being a subalgebra of $J_{\alpha+1}/J_\alpha$, $I_{\alpha+1}/I_\alpha$ is CCR. Thus I is GCR.

Next we will show that $\{J_\alpha + I/I : 0 \leq \alpha \leq \alpha_0\}$ is a composition series for A/I. Let $\beta < \alpha_0$ be a limit ordinal. Clearly, $J_\beta + I \subseteq$ norm closure of $\bigcup_{\alpha<\beta} J_\alpha + I$. Now, $\bigcup_{\alpha<\beta} J_\alpha + I \subset J_\beta + I$. Therefore, $\bigcup_{\alpha<\beta} J_\alpha + I \subset J_\beta + I$, since $J_\beta + I$ being the sum of two closed ideals in a C^*-algebra is closed (see [3]). Hence, norm closure of $\bigcup_{\alpha<\beta} J_\alpha + I = J_{\beta+1}$. Thus $\{J_\alpha + I/I : 0 \leq \alpha \leq \alpha_0\}$ is a composition series for A/I.

Also $J_{\alpha+1} + I/I/J_\alpha/I \cong J_{\alpha+1} + I/J_\alpha + I \cong J_{\alpha+1}/J_{\alpha+1} \cap J_\alpha + I \subseteq J_{\alpha+1}/J_\alpha$. Therefore $J_{\alpha+1} + I/I/J_\alpha + I/I$ is CCR. Hence A/I is GCR. □

Remark 3.6 In the previous theorem, if we start with the unique composition series $\{J_\alpha : 0 \leq \alpha \leq \alpha_0\}$ for A, then we can show that the composition series $\{J_\alpha \cap I : 0 \leq \alpha \leq \alpha_1\}$and $\{J_\alpha + I/I : 0 \leq \alpha \leq \alpha_0\}$ are the unique composition series for I and A/I, respectively, as follows.

Let $x_1 \in J_\alpha \cap I$ and π be an irreducible representation of $I/J_\alpha \cap I$. We can identify $I/J_\alpha \cap I$ with an ideal in A/J_α, by means of the function $\phi : I/J_\alpha \cap I \mapsto A/J_\alpha$ given by $\phi(x + (J_\alpha \cap I)) = x + J_\alpha$. Now, let π_1 be the irreducible extension of π to A/J_α. Then, $\pi(x_1 + (J_\alpha \cap I)) = \pi_1(x_1 + J_\alpha)$ is compact since $J_{\alpha+1}/J\alpha =$ CCR(A/J_α). Hence $J_{\alpha+1} \cap I/J_\alpha \cap I = CCR(I/J_\alpha)$.

Now, let $x_2 \in J_{\alpha+1} + I$ and π be an irreducible representation of $A/(J_\alpha + I)$. Define an irreducible representation π_1 of A/J_α by letting $\pi_1(x + J_\alpha) = \pi(y + (J_\alpha + I))$.

Then, $\pi(x_2 + (J_\alpha + I)) = \pi(x_2 + J_\alpha)$ is compact since $J_{\alpha+1}/J_\alpha =$ CCR(A/J_α). Therefore $J_{\alpha+1} + I/J_\alpha + I = CCR(A/J_\alpha + I)$.

Acknowledgments This article is based on a joint work with Ms. A.M. Shabna who was CSIR research fellow at the National Institute of Technology Calicut during the work.

References

1. Arverson, W.: Invitation to C^*-algebra. Graduate texts in Mathematics. Springer, Berlin (1976)
2. Castillo, J.M.F., Gonzalez, M.: Three-space problems in Banach space theory. Lecture Notes in Mathematics, vol. 1667. Springer, Berlin (1997)
3. Dixmier, J.: C^*-algebras. North-Holland, New York (1982)
4. Enflo, P., Lindenstrauss, J., Pisier, G.: On the three-space problem. Math. Scand. **36**, 199–210 (1975)
5. Godefroy, G., Troyanski, S., Whitefield, J., Zizler, V.: Three-space problem for locally uniformly rotund renormings of Banach spaces. Proc. Am. Math. Soc. **94**, 647–652 (1985)
6. Kusuda, M.: Three-space problems in discrete spectra of C^*-algebras and dual C^*-algebras. Proc. R. Soc. Edinb. **131A**, 701–707 (2001)
7. Krein, M., Smullian, V.: On regularly convex sets in the space conjugate to a Banach space. Ann. Math. **41**, 556–583 (1940)

8. Sanchez, F.C., Castillo, J.M.F.: Duality and twisted sums of Banach spaces. J. Funct. Anal. **175**, 1–16 (2000)
9. Rao, T.S.S.R.K.: Three-space problem for some classes of L^1-preduals, J. Math. Anal. Appl. **351**, 311–314 (2009)

Spectral Approximation of Bounded Self-Adjoint Operators—A Short Survey

K. Kumar

Abstract Normal categories are essentially those arising as the category of principal left [right] ideals of a regular semigroup. These categories have been used in describing the structure of regular semigroups. The structure theory in this context is known as cross connection theory. Several associated categories can be derived from a normal category which are also of interest in the structure theory of regular semigroups. The subcategory of inclusions, the subcategory of retractons, the groupoid of isomorphisms etc. are some of the associated categories.

Keywords Self-adjoint operators · Spectrum · Truncation · Filteration of a Hilbert space · Arveson's class operator · Essential spectrum · Gaps in spectrum

Mathematics Subject Classification (2010) Primary 47B80 · Secondary 47H40 · 47B15

1 Introduction

The fundamental question "How to approximate spectra of linear operators on separable Hilbert spaces?" was considered by many mathematicians, starting from Szegö in [21]. Several attempts have been made to make use of the finite dimensional theory

An erratum of this chapter can be found under DOI 10.1007/978-81-322-2488-4_17

K. Kumar (✉)
Department of Mathematics, PSMO College, Tirurangadi 676306, Kerala, India
e-mail: kiranbalu36@gmail.com

P.G. Romeo et al. (eds.), *Semigroups, Algebras and Operator Theory*,
Springer Proceedings in Mathematics & Statistics 142,
DOI 10.1007/978-81-322-2488-4_15

in the computation of the spectrum of bounded operators in an infinite dimensional space through an asymptotic way. This approach found success in getting good estimates in the case of some self-adjoint operators. Significant efforts have been done by many mathematicians to build up a general theory for the approximation of the spectrum of bounded self-adjoint operators on an infinite dimensional Hilbert space. To quote some of the recent contributions in this direction are due to W.B. Arveson [1], Albrecht Böttcher et al. [4], E.B. Davies et al. [6, 7], I. Gohberg et al. [9], A. Hansen [11], etc. The list is nevertheless incomplete.

A short survey is presented here on various techniques used to approximate the spectrum of a bounded self-adjoint operator A on a separable complex Hilbert space $\mathbb{H}$. The finite dimensional compressions A_n of A are considered here. The asymptotic values of spectrum of A_n are used to study the nature of spectrum of A.

1.1 The Problem

Let $\{e_1, e_2, \ldots\}$ be an orthonormal basis for $\mathbb{H}$ and P_n be the projection of $\mathbb{H}$ onto the finite dimensional subspace $L_n = \text{span}\{e_1, e_2, \ldots, e_n\}$. The finite dimensional truncations $A_n = P_n A P_n$ of A can be treated as finite matrices by restricting their domains to the image of P_n. If we denote the infinite matrix $(a_{i,j}) = (\langle Ae_j, e_i\rangle)$ to be the matrix representation of A associated to the orthonormal basis $\{e_1, e_2, \ldots\}$, then the $n \times n$ matrix $(a_{i,j})_{1\le i,j\le n}$ coincides with the matrix representation of A_n restricted to the image of P_n.

Here we consider the following fundamental question. Can we approximate the spectrum of A using the eigenvalue sequences of the matrices $(a_{i,j})_{1\le i,j\le n}$. There are some disappointing examples in which the eigenvalues of truncations give little information about the spectrum. For instance, in the case of the right shift operator on the sequence space $l^2(\mathbb{Z})$, the eigenvalue sequence of the truncations is the constant sequence 0, while the spectrum is the whole closed unit circle. For a self-adjoint example, one can consider the operator A on $l^2(\mathbb{N})$, defined as follows:

$$A(x_n) = (x_{\pi(n)}), \tag{1.1}$$

where π is a suitably chosen permutation on $\mathbb{N}$. The essential properties required for the permutation π, are discussed in [1], due to which the truncation method fails to approximate the spectrum.

This article is a survey of some recent developments in this area. In the next section, we discuss the class of operators introduced by W.B. Arveson in [1] for which the spectrum is fully determined by the eigenvalues of their truncations except for some discrete eigenvalues that may lie between the bounds of essential spectrum. Also, the use of the truncation method to approximate the bounds and the discrete eigenvalues lying outside the bounds of the essential spectrum of a bounded self-adjoint operator is explained in this section. The recent advances in the spectral gap prediction problems are also discussed there. The use of preconditioners to modify the truncation method is explained with a couple of more recent results. In the third section, we briefly explain the quadratic projection method and second-order relative spectra with some recent modifications. A concluding section on the further possibilities ends the article.

2 Algebraic and Linear Algebraic Approach

First we report some of the algebraic developments in this area. The major contributions are due to W.B. Arveson, who generalized the notion of band-limited matrices in [1], and achieved some success in the case of a special class of operators. We start with some definitions and results below which will play a very important role in the approximation of the spectrum of bounded self-adjoint operators. The notation A_n is used to denote the matrix $(a_{i,j})_{1\leq i,j\leq n}$.

Definition 2.1 A **filtration** of a Hilbert space $\mathbb{H}$ is a sequence of finite dimensional subspaces of $\mathbb{H}$, $\{L_n; n \in \mathbb{N}\}$ such that $L_n \subset L_{n+1}$ and closure of the union $\bigcup_n L_n$ is $\mathbb{H}$.

Example 2.2 A typical example for filtration in a Hilbert space with an orthonormal basis is the following. Let $\{e_n : n \in \mathbb{Z}\}$ be the bilateral orthonormal basis for $\mathbb{H}$ and let $\{L_n\}$ be defined by $L_n = \text{span}\{e_{-n}, e_{-n+1}, \dots e_n\}$. Then $\{L_n; n \in \mathbb{Z}\}$ is a filtration.

Definition 2.3 Let $\{L_n : n \in \mathbb{N}\}$ be a filtration. And P_n be the projection onto L_n. The **degree** of a bounded operator A on $\mathbb{H}$ is defined by

$$\deg(A) = \sup_{n\geq 1} \text{rank}(P_n A - A P_n).$$

Corresponding to each filtration, a Banach $*-$algebra of operators called Arveson's class can be defined as follows.

Definition 2.4 A is an operator in the **Arveson's class** if $A = \sum_{n=1}^{\infty} A_n$, where $\deg(A_n) < \infty$ for every n and convergence is in the operator norm, in such a way that $\sum_{n=1}^{\infty}(1 + \deg(A_n)^{\frac{1}{2}})\|A_n\| < \infty$.

In case each L_n is the span of finite number of elements in the basis as defined in Example 2.2, the following gives a concrete description of operators in the Arveson's class.

Theorem 2.5 ([1]) *Let $\{L_n; n \in \mathbb{Z}\}$ be the filtration defined in Example 2.2. Also let $(a_{i,j})$ be the matrix representation of a bounded operator A, with respect to $\{e_n\}$, and for every $k \in \mathbb{Z}$ let*

$$d_k = \sup_{i\in\mathbb{Z}} |a_{i+k,i}|$$

be the sup norm of the kth diagonal of $(a_{i,j})$. *Then A will be in the Arveson's class whenever the series* $\sum_k |k|^{1/2} d_k$ *converges.*

In particular, any operator whose matrix representation $(a_{i,j})$ is band-limited, in the sense that $a_{i,j} = 0$ whenever $|i - j|$ is sufficiently large, must be in the Arveson's class. Before stating the spectral inclusion theorems for arbitrary self-adjoint operators and for operators in the Arveson's class, recall the notion of essential points and transient points.

Definition 2.6 Essential point: A real number λ is an essential point of A, if for every open set U containing λ, $\lim_{n\to\infty} N_n(U) = \infty$, where $N_n(U)$ is the number of eigenvalues of A_n in U.

Definition 2.7 Transient point: A real number λ is a transient point of A if there is an open set U containing λ, such that $\sup N_n(U)$ with n varying on the set of all natural number, is finite.

Remark 2.8 It should be noted that a number can be neither transient nor essential.

Denote $\Lambda = \{\lambda \in R; \lambda = \lim \lambda_n, \lambda_n \in \sigma(A_n)\}$ and Λ_e as the set of all essential points. The following spectral inclusion results for a bounded self-adjoint operator A is of high importance.

Theorem 2.9 ([1]) *The spectrum of a bounded self-adjoint operator is contained in the set of all limit points of the eigenvalue sequences of its truncations. Also, the essential spectrum is contained in the set of all essential points, i.e.,*

$$\sigma(A) \subseteq \Lambda \subseteq [m, M] \;\; and \;\; \sigma_e(A) \subseteq \Lambda_e.$$

Equality in one of the above inclusion for bounded self-adjoint operators in the Arveson's class, was also proved in [1]. The precise result is the following.

Theorem 2.10 ([1]) *If A is a bounded self-adjoint operator in the Arveson's class, then* $\sigma_e(A) = \Lambda_e$ *and every point in* Λ *is either transient or essential.*

Remark 2.11 The above two theorems enable us to confine our attention to the limiting set Λ and the essential points Λ_e, in the task of computation of spectrum and essential spectrum of a bounded self-adjoint operator, respectively. Now the following issues may arise. The limiting set Λ may contain points which does not belong to the spectrum. Such points are called spurious eigenvalues. In the case of an operator in the Arveson's class, the essential points will give all information about essential spectrum, while the transient points may be misleading. Here we loose only information about eigenvalues of finite multiplicity. But this is very important if such points exist between the lower and upper bounds of essential spectrum, since they lead to the existence of spectral gaps between these bounds.

2.1 Operators with Connected Essential Spectrum

Things can be more difficult in the case of an arbitrary bounded self-adjoint operator. There may exist essential points, which are not spectral values. The operator given by the Eq. (1.1) is of that kind. However, the inclusion in Theorem 2.9 helps us to determine the spectrum, with an additional assumption of connectedness of the essential spectrum. The details of this claim are given below, which is a brief review of the article [4] with some slight modifications. This will play a key role in the forthcoming sections.

Recall that, for a bounded self-adjoint operator A, the spectrum $\sigma(A)$ is contained in the interval $[m, M]$ and the essential spectrum $\sigma_e(A)$ in $[\nu, \mu]$, where m, M, ν, μ, are bounds of $\sigma(A)$ and $\sigma_e(A)$, respectively. The following definitions and preliminary results are needed further.

Definition 2.12 Consider the singular number s_k, k natural number,

$$s_k(A) = \inf\{\|A - F\|\,;\, F \in \mathbb{B}(\mathbb{H}), \text{rank } F \leq k-1\}$$

is the kth **approximation number** of A.

Clearly, we have $\|A\| = s_1(A) \geq s_2(A) \geq \cdots \geq 0$

Theorem 2.13

- [9] $\lim_{k\to\infty} s_k(A) = \|A\|_{ess}$ *where* $\|A\|_{ess}$ *is the essential norm.*
- [4] $\lim_{n\to\infty} s_k(A_n) = s_k(A)$.

Remark 2.14 For $|A| = (A^*A)^{1/2}$, in case A is a finite matrix, the approximation numbers are the eigenvalues of $|A|$. That is $s_k(A) = \lambda_k(|A|)$, where $\lambda_k(|A|)$ is the kth eigenvalue of $|A|$.

Theorem 2.15 ([9]) *The set* $\sigma(|A|) - [0, \|A\|_{ess}]$ *is at most countable,* $\|A\|_{ess}$ *is the only possible accumulation point, and all the points of the set are eigenvalues with finite multiplicity of* $|A|$. *Furthermore if*

$$\lambda_1(|A|) \geq \lambda_2(|A|) \geq \cdots \geq \lambda_N(|A|)$$

are those N eigenvalues (N can be infinity), then

$$s_k(A) = \begin{cases} \lambda_k(|A|), & \text{if } N = \infty \text{ or } 1 \leq k \leq N \\ \|A\|_{ess}, & \text{if } N < \infty \text{ and } k \geq N+1 \end{cases} \tag{2.1}$$

Corollary 2.16

$$\lim_{n\to\infty} \lambda_k(|A_n|) = \lim_{n\to\infty} s_k(A_n) = s_k(A) = \begin{cases} \lambda_k(|A|) \text{ if } N = \infty \text{ or } 1 \leq k \leq N \\ \|A\|_{ess} \text{ if } N < \infty \text{ and } k \geq N+1 \end{cases}$$

Remark 2.17 The above result will play a key role in the approximation of spectrum. Considering the positive operator $A - mI$, it can be deduced that the set $\sigma(A) \cap (\mu, M]$ is at most countable and that consists of eigenvalues of finite multiplicity by Theorem 2.15. Also μ is the only possible accumulation point. Let these eigenvalues be

$$\lambda_R^+(A) \leq \cdots \leq \lambda_2^+(A) \leq \lambda_1^+(A).$$

Similarly by considering the operator $MI - A$, it can be observed that $\sigma(A) \cap [m, \nu)$ consists of at most countably many eigenvalues of finite multiplicity with only possible accumulation point ν. Let

$$\lambda_1^-(A) \leq \lambda_2^-(A) \leq \cdots \leq \lambda_S^-(A)$$

be those eigenvalues. Also the numbers R and S can be infinity. Arrange the eigenvalues of A_n as

$$\lambda_1(A_n) \geq \lambda_2(A_n) \geq \cdots \geq \lambda_n(A_n).$$

From here onwards, the above notations will be used.

Now we prove the following result from [4] which is the major tool that is used frequently in this note.

Theorem 2.18 *For every fixed integer k we have*

$$\lim_{n\to\infty} \lambda_k(A_n) = \begin{cases} \lambda_k^+(A), & \text{if } R = \infty \text{ or } 1 \leq k \leq R \\ \mu, & \text{if } R < \infty \text{ and } k \geq R+1 \end{cases}$$

$$\lim_{n\to\infty} \lambda_{n+1-k}(A_n) = \begin{cases} \lambda_k^-(A), & \text{if } S = \infty \text{ or } 1 \leq k \leq S \\ \nu, & \text{if } S < \infty \text{ and } k \geq S+1 \end{cases}$$

In particular,

$$\lim_{k\to\infty}\lim_{n\to\infty} \lambda_k(A_n) = \mu \text{ and } \lim_{k\to\infty}\lim_{n\to\infty} \lambda_{n+1-k}(A_n) = \nu.$$

Proof The following observations are made first.

$$|A - mI| = A - mI, \; P_n(A - mI)P_n = A_n - mI_n, \text{ and } |A_n - mI_n| = A_n - mI_n.$$

Hence from the above corollary, we have

$$\lim_{n\to\infty} \lambda_k(A_n - mI_n) = \begin{cases} \lambda_k(A - mI), & \text{if } R = \infty \text{ or } 1 \leq k \leq R \\ \|A - mI\|_{ess}, & \text{if } R < \infty \text{ and } k \geq R+1 \end{cases} \tag{2.2}$$

Similarly, by considering the operator $MI - A$, we get

$$\lim_{n\to\infty} \lambda_k(MI_n - A_n) = \begin{cases} \lambda_k(MI - A), & \text{if } S = \infty \text{ or } 1 \le k \le S \\ \|MI - A\|_{ess}, & \text{if } S < \infty \text{ and } k \ge S+1 \end{cases} \tag{2.3}$$

Also we have the following identities

$$\|A - mI\|_{ess} = \mu - m, \quad \|MI - A\|_{ess} = M - \nu. \tag{2.4}$$

$$\lambda_k(A_n - mI_n) = \lambda_k(A_n) - m, \quad \lambda_k(MI_n - A_n) = M - \lambda_{n+1-k}(A_n). \tag{2.5}$$

$$\lambda_k(A - mI) = \lambda_k^+(A) - m, \quad \lambda_k(MI - A) = M - \lambda_k^-(A). \tag{2.6}$$

Substituting them in Eqs. (2.2) and (2.3), we get

$$\lim_{n\to\infty} \lambda_k(A_n) = \begin{cases} \lambda_k^+(A), & \text{if } R = \infty \text{ or } 1 \le k \le R \\ \mu, & \text{if } R < \infty \text{ and } k \ge R+1 \end{cases}$$

$$\lim_{n\to\infty} \lambda_{n+1-k}(A_n) = \begin{cases} \lambda_k^-(A), & \text{if } S = \infty \text{ or } 1 \le k \le S \\ \nu, & \text{if } S < \infty \text{ and } k \ge S+1 \end{cases}$$

Hence the proof. □

Remark 2.19 The above results are also true if we replace A_n by some other sequence A_{1n} of self-adjoint operators with the property that

$$\|A_n - A_{1n}\| \to 0 \text{ as } n \to \infty$$

In order to justify this, we need only to recall an important inequality concerning the eigenvalues of self-adjoint matrices A, B (refer e.g. to [2])

$$|\lambda_k(A) - \lambda_k(B)| \le \|A - B\|.$$

Remark 2.20 By Theorem 2.18, all the discrete spectral values lying outside the bounds of essential spectrum and the upper and lower bounds of the essential spectrum can be approximated. Note that, the theorem points out exactly the particular sequence that converges to a discrete spectral value. But how fast does the convergence take place, is still not known. Looking at some concrete situations, one may hope for a better rate of convergence.

Even the rate of convergence is not estimated, it can be proved that the order of convergence is the same as the order of convergence of approximation numbers. The following theorem gives a vague idea about the rate of convergence.

Theorem 2.21 ([14]) *If $s_k(A_n) - s_k(A) = O(\theta_n)$, where θ_n goes to 0 as n tends to ∞, then*

$$\lambda_k(A_n) = \begin{cases} \lambda_k^+(A) + O(\theta_n), & \textit{if } R = \infty \textit{ or } 1 \leq k \leq R \\ \mu + O(\theta_n), & \textit{if } R < \infty \textit{ and } k \geq R+1 \end{cases}$$

$$\lambda_{n+1-k}(A_n) = \begin{cases} \lambda_k^-(A) + O(\theta_n), & \textit{if } S = \infty \textit{ or } 1 \leq k \leq S \\ \nu + O(\theta_n), & \textit{if } S < \infty \textit{ and } k \geq S+1 \end{cases}$$

where R and S are the same notations used in Theorem 2.18.

Proof Let N be the number of eigenvalues lying in $\sigma(|A|) - \left[0, \|A\|_{ess}\right]$. From identity (2.1), and the the fact that $s_k(A_n) = \lambda_k(|A_n|)$, we have the following identity.

$$s_k(A_n) - s_k(A) = \begin{cases} \lambda_k(|A_n|) - \lambda_k(|A|), & \text{if } N = \infty \text{ or } 1 \leq k \leq N \\ \lambda_k(|A_n|) - \|A\|_{ess}, & \text{if } N < \infty \text{ and } k \geq N+1 \end{cases}$$

Since by hypothesis, $s_k(A_n) - s_k(A) = O(\theta_n)$,

$$\lambda_k(|A_n|) - \lambda_k(|A|) = O(\theta_n), \quad \text{if } N = \infty \text{ or } 1 \leq k \leq N,$$

$$\lambda_k(|A_n|) - \|A\|_{ess} = O(\theta_n), \quad \text{if } N < \infty \text{ and } k \geq N+1.$$

Applying this to the positive operators $A - mI$, and $MI - A$, with the notations used in Theorem 2.18, we get the following conclusions.

$$\lambda_k(A_n - mI_n) = \begin{cases} \lambda_k(A - mI) + O(\theta_n), & \text{if } R = \infty \text{ or } 1 \leq k \leq R \\ \|A - mI\|_{ess} + O(\theta_n), & \text{if } R < \infty \text{ and } k \geq R+1 \end{cases}$$

and

$$\lambda_k(MI_n - A_n) = \begin{cases} \lambda_k(MI - A) + O(\theta_n), & \text{if } S = \infty \text{ or } 1 \leq k \leq S \\ \|MI - A\|_{ess} + O(\theta_n), & \text{if } S < \infty \text{ and } k \geq S+1 \end{cases}$$

Also from the identities (2.4)–(2.6), we get the desired conclusions

$$\lambda_k(A_n) = \begin{cases} \lambda_k^+(A) + O(\theta_n), & \text{if } R = \infty \text{ or } 1 \leq k \leq R \\ \mu + O(\theta_n), & \text{if } R < \infty \text{ and } k \geq R+1 \end{cases}$$

$$\lambda_{n+1-k}(A_n) = \begin{cases} \lambda_k^-(A) + O(\theta_n), & \text{if } S = \infty \text{ or } 1 \leq k \leq S \\ \nu + O(\theta_n), & \text{if } S < \infty \text{ and } k \geq S+1 \end{cases}$$

Hence the proof. □

The above theorem is the first result regarding the rate of convergence in the approximations done in Theorem 2.18. So far there is no evidence of remainder estimation and the error estimation in these approximations in the case of an arbitrary self-adjoint operator to the best of our knowledge. The subsequent theorem taken from [4] denies the existence of spurious eigenvalues (points in Λ those are not part of the spectrum) under the assumption of connectedness of essential spectrum.

Theorem 2.22 ([4]) *If A is a self-adjoint operator and if $\sigma_e(A)$ is connected, then $\sigma(A) = \Lambda$.*

Remark 2.23 It is worthwhile to notice that the connectedness of essential spectrum enables us to compute the spectrum using finite dimensional truncations. Thus, if we cannot determine the spectrum fully by the truncations, then the essential spectrum is not connected. In short, if there is a spurious eigenvalue, then there exists a gap in the essential spectrum.

Remark 2.24 The converse of the above observation need not be true. That is the existence of a spectral gap does not lead to the existence of a spurious eigenvalue. For example, if we take A to be be the projection operator on to some closed subspace of $\mathbb{H}$, then the eigenvalues of truncations are 0 and 1 only. There we have $\Lambda = \sigma(A) = \{0, 1\}$. Hence no spurious eigenvalues, but still there is a gap.

In summary, the upper and lower bounds of the essential spectrum can be computed using the sequence of eigenvalues of finite dimensional truncations. Also the discrete eigenvalues lying below and above these bounds can be computed. The above results pinpointing the particular sequence of eigenvalues that converges to a particular eigenvalue of the operator. Now the remaining part is the computation of essential spectrum. The problem is whether it is possible to locate the gaps in the essential spectrum using these truncations. If it is possible, then the spectrum is fully determined up to some discrete eigenvalues that may have trapped between these gaps.

2.2 Gaps in the Essential Spectrum

The following theorem is an attempt to predict the existence of spectral gaps, using the finite dimensional truncations. The notation $\#S$ is used to denote the number of elements in the set S and w_{nk} is used to denote an averaging sequence. That is $0 \leq w_{nk} \leq 1$, and $\sum_{k=1}^{n} w_{nk} = 1$.

Theorem 2.25 ([13]) *Let A be a bounded self-adjoint operator, and $\lambda_{n1}(A_n) \geq \lambda_{n2}(A_n) \geq \cdots \geq \lambda_{nn}(A_n)$ be the eigenvalues of A_n arranged in decreasing order. For each positive integer n, let $a_n = \sum_{k=1}^{n} w_{nk}\lambda_{nk}$ be the convex combination of eigenvalues of A_n. If there exists a $\delta > 0$ and $K > 0$ such that*

$$\#\left\{\lambda_{nj};\, \left|a_n - \lambda_{nj}\right| < \delta\right\} < K \tag{2.7}$$

and in addition if $\sigma_e(A)$ and $\sigma(A)$ have the same upper and lower bounds, then $\sigma_e(A)$ has a gap.

Remark 2.26 There is possibility for the presence of discrete eigenvalues inside the gaps in the above case.

Remark 2.27 The special case which is more interesting is when $w_{nk} = \frac{1}{n}$, for all n. In that case, we are actually looking at the averages of eigenvalues of truncations and these averages can be computed using the trace at each level.

Remark 2.28 It is to be noted that all the points of the form $a_n = \sum_{k=1}^{n} w_{nk}\lambda_{nk}$ are in the numerical range of A_n. Therefore, the result can be made simpler in the language of numerical range. However it is not easy to compute the numbers in the expression (2.7). Here we treated it as a deviation from the mean value. Hence the condition (2.7) may be interpreted as a restriction to the deviation of the eigenvalues of truncations from their central tendency. Nevertheless the computations still remain difficult.

In Theorem 2.25, the weighted mean of the eigenvalues at each level and its deviation is analyzed. The following special choice of the weights are interesting.

Special Choice

Let us consider an instance where these weights w_{nk} arise naturally associated to a self-adjoint operator on a Hilbert space. Let $A_n = \sum_{k=1}^{n} \lambda_{n,k} Q_{n,k}$ be the spectral resolution of A_n. Define $w_{nk} = \langle Q_{n,k}e_1, e_1\rangle$. Then $0 \leq w_{nk} \leq 1$ and $\sum_{k=1}^{n} w_{nk} = 1$. Now

$$\sum_{k=1}^{n} w_{nk}\lambda_{nk} = \sum_{k=1}^{n} \lambda_{nk}\langle Q_{n,k}e_1, e_1\rangle = \langle A_n e_1, e_1\rangle = \langle Ae_1, e_1\rangle = a_{11}.$$

Therefore by Theorem 2.25, if there exists a $\delta > 0$ and a $K > 0$, such that

$$\#\left\{\lambda_{nj};\ |a_{11} - \lambda_{nj}| < \delta\right\} < K$$

then there exists a gap in the essential spectrum of A. Hence if the first entry in the matrix representation of A, is not an essential point, then there exists a gap in the essential spectrum.

Remark 2.29 All points of the form $\langle Ae_i, e_i\rangle = a_{ii}$ are in the numerical range which lies between the bounds of the essential spectrum, in the case that the bounds coincide with the bounds of the spectrum. Hence in that case, if a_{ii} is not an essential point for some i, then that will lead to the existence of a spectral gap. That means if any one of the diagonal entries in the matrix representation of A is not an essential point, then there exists a gap in the essential spectrum as indicated in the above special choice of w_{nk}.

The following is an example where the first entry a_{11} is a transient point and the spectral gap prediction is valid.

Example 2.30 Define a bounded self-adjoint operator A on $l^2(\mathbb{N})$, as follows.

$$A(x_n) = (x_{n-1} + x_{n+1}) + (v_n x_n), x_0 = 0;$$

where the periodic sequence $(v_n) = (1, 2, 3, 1, 2, 3, \ldots)$. The matrix representation of A, associated to the standard orthonormal basis, is tridiagonal. The diagonal entries are the entries in the periodic sequence (v_n) and upper and lower diagonal will be 1. Such matrices can be identified as the block Toeplitz operator with corresponding matrix valued symbol given by

$$\tilde{f}(\theta) = \begin{bmatrix} 1 & 1 & e^{i\theta} \\ 1 & 2 & 1 \\ e^{-i\theta} & 1 & 3 \end{bmatrix}.$$

By our special choice above, Theorem 2.25 guarantees that if $\langle A(e_1), e_1 \rangle = 1$ is a transient point, then $\sigma_e(A)$ has a gap. The fact that 1 is a transient point, is a consequence of discrete Borg theorem [8, 10] and some numerical computations. The interval $\left(\frac{3-\sqrt{5}}{2}, \frac{5-\sqrt{5}}{2}\right)$ is a spectral gap an 1 lies in that gap.

2.3 Preconditioners in Spectral Approximation

Here we try to modify the truncation method with the help of the notions of preconditioners and the convergence of matrix sequences in the sense of eigenvalue clustering. Recall that in the numerical analysis literature, the preconditioner associated with a matrix is used to make the iteration process more efficient. Here we use different notions of matrix convergence in the sense of eigenvalue clustering to study the spectral approximation by preconditioners. That is, the A_n's will be replaced by its preconditioner to perform approximation of spectrum.

We start with defining different notions of convergence of matrix sequences in the sense of eigenvalue clustering. Such notions were used in the special case of Toeplitz matrices in [20], and generalized into the arbitrary case in [12].

Definition 2.31 Let $\{A_n\}$ and $\{B_n\}$ be two sequences of $n \times n$ Hermitian matrices. We say that $A_n - B_n$ converges to 0 in the ***strong cluster sense*** if for any $\epsilon > 0$, there exist integers $N_{1,\epsilon}, N_{2,\epsilon}$ such that all the singular values $\sigma_j(A_n - B_n)$ lie in the interval $[0, \epsilon)$ except for at most $N_{1,\epsilon}$ (independent of the size n) singular values for all $n > N_{2,\epsilon}$.

If the number $N_{1,\epsilon}$ does not depend on ϵ, we say that $A_n - B_n$ converges to 0 in the ***uniform cluster sense***. And if $N_{1,\epsilon}$ depends on ϵ, n and is of $o(n)$, we say that $A_n - B_n$ converges to 0 in ***weak cluster sense***.

Here the aim is to modify the truncation method by replacing A_n by some other *simpler* sequence of matrices B_n, where $\{A_n\} - \{B_n\}$ converges to 0 in the strong cluster sense (weak or uniform cluster sense, respectively). We study the effect of this replacement in the well-known results obtained by truncation method. We prove a couple of results which show that the convergence in the strong or uniform cluster sense is equivalent to the compact perturbation of operators. These are the modified versions of the results proved in [15].

Theorem 2.32 *Let $A, B \in B(\mathcal{H})$ be self-adjoint operators. Then the operator $R = A - B$ is compact if and only if the sequence of truncations $A_n - B_n$ converges to the zero matrix in the strong cluster.*

Proof First assume that $R = A - B$ is compact and its spectrum $\sigma(R) = \{\lambda_k(R) : k = 1, 2, 3, \ldots\} \bigcup \{0\}$. Here 0 is the only accumulation point of the spectrum. Hence $\lambda_k(R) \to 0$ as $k \to \infty$. Hence for any given $\epsilon > 0$, there exists a positive integer $N_{1,\epsilon}$ such that

$$\lambda_k(R) \in \left(\frac{-\epsilon}{2}, \frac{\epsilon}{2}\right), \text{ for every } k > N_{1,\epsilon}.$$

Also since R is compact, the truncation $R_n = A_n - B_n$ converges to R in the operator norm topology. Therefore, the eigenvalues of truncations converges to the eigenvalues of R. That is

$$\lambda_k(R_n) \to \lambda_k(R) \text{ as } n \to \infty, \text{ for each } k.$$

In particular, for every $k > N_{1,\epsilon}$, there exists a positive integer $N_{2,\epsilon}$ such that

$$\lambda_k(R_n) - \lambda_k(R) \in \left(\frac{-\epsilon}{2}, \frac{\epsilon}{2}\right), \text{ for every } n > N_{2,\epsilon}.$$

Therefore, when $n > N_{2,\epsilon}$, all the eigenvalues $\lambda_k(R_n)$ of $R_n = A_n - B_n$, except for at most $N_{1,\epsilon}$ eigenvalues, are in the interval $(-\epsilon, \epsilon)$. That is $R_n = A_n - B_n$ converges to 0 in the strong cluster.

For the converse part, assume that $A_n - B_n$ converges to the zero matrix in the strong cluster. Then for any $\lambda \neq 0$, choose an $\epsilon > 0$ such that λ is outside the interval $(-\epsilon, \epsilon)$. Corresponding to this ϵ, there exist positive integers $N_{1,\epsilon}, N_{2,\epsilon}$ such that $\sigma(A_n - B_n)$ is contained in $(-\epsilon, \epsilon)$, for every $n > N_{2,\epsilon}$, except for possibly $N_{1,\epsilon}$ eigenvalues. Now consider the counting function $N_n(U)$ of eigenvalues of $A_n - B_n$ in U. For any neighborhood U of λ that does not intersect with $(-\epsilon, \epsilon)$, $N_n(U)$ is bounded by the number $N_{1,\epsilon}$. Hence λ is not an essential point of $A - B$. Therefore, it is not in the essential spectrum (see Theorem 2.3 of [1]). Since $\lambda \neq 0$ was arbitrary, this shows that the essential spectrum of $A - B$ is the singleton set $\{0\}$. Hence it is a compact operator and the proof is completed. □

Theorem 2.33 *Let $A, B \in B(\mathcal{H})$ be self-adjoint operators. Then the operator $R = A - B$ is of finite rank if and only if the truncations $A_n - B_n$ converges to the zero matrix in the uniform cluster.*

Proof The proof is an imitation of the proof of Theorem 2.32, differs only in the choice of $N_{1,\epsilon}$ to be independent of ϵ. However the details are given below. First assume that R = A − B is a finite rank operator with rank N_1, and its spectrum $\sigma(R) = \{\lambda_k(R) : k = 1, 2, 3, \dots N_1\} \bigcup \{0\}$. Since the truncation $R_n = A_n - B_n$ converges to R in the operator norm topology, the eigenvalues of truncations converges to the eigenvalues of R. That is

$$\lambda_k(R_n) \to \lambda_k(R) \text{ as } n \to \infty, \text{ for each } k = 1, 2, 3, \dots N_1.$$

For every $k > N_1$, $\lambda_k(R_n)$ converges to 0 by [4]. Hence for a given $\epsilon > 0$, there exists a positive integer $N_{2,\epsilon}$ such that

$$\lambda_k(R_n) \in (-\epsilon, \epsilon), \text{ for every } n > N_{2,\epsilon} \text{ and for each } k > N_1.$$

Therefore, when $n > N_{2,\epsilon}$, all the eigenvalues $\lambda_k(R_n)$ of $R_n = A_n - B_n$, except for the first N_1 eigenvalues, are in the interval $(-\epsilon, \epsilon)$. That is $A_n - B_n$ converges to 0 in the uniform cluster.

For the converse part, assume that $A_n - B_n$ converges to the zero matrix in the uniform cluster. Then for any $\epsilon > 0$, there exist positive integers N_1, $N_{2,\epsilon}$ such that $\sigma(A_n - B_n)$ is contained in $(-\epsilon, \epsilon)$, for every $n > N_{2,\epsilon}$, except for possibly N_1 eigenvalues. As in the proof of Theorem 2.32, we obtain 0 is the only element in the essential spectrum. Hence $R = A - B$ is a compact operator. In addition to this, R can have at most N_1 eigenvalues. To see this, notice that all the eigenvalues of a compact operator are obtained as the limits of sequence of eigenvalues of its truncations. In this case at most N_1 such sequence can go to a nonzero limit. Hence R is a finite rank operator and the proof is completed. □

Remark 2.34 The above results have the following implications. Since a compact perturbation may change the discrete eigenvalues, the above results show that the convergence of preconditioners in the sense of eigenvalue clustering, is not sufficient to use them in the spectral approximation problems. Nevertheless one can use it in the spectral gap prediction problems, since the compact perturbation preserves essential spectrum.

Remark 2.35 The analysis of weak convergence is yet to be carried out.

We end this section with the example of Frobenius optimal preconditioners, which are useful in the context of infinite linear systems with Toeplitz structure (see [20] for details).

Example 2.36 Let $\{U_n\}$ be a sequence of unitary matrices over $\mathbb{C}$, where U_n is of order n for each n. For each n, we define the commutative algebra M_{U_n} of matrices as follows.

$$M_{U_n} = \left\{A \in M_n(\mathbb{C}) \, ; \, {U_n}^* A U_n \text{ complex diagonal}\right\}$$

Recall that $M_n(\mathbb{C})$ is a Hilbert space with respect to the classical Frobenius scalar product,

$$\langle A, B\rangle = \text{trace } (B^* A).$$

Observe that M_{U_n} is a closed convex set in $M_n(\mathbb{C})$ and hence, corresponding to each $A \in M_n(\mathbb{C})$, there exists a unique matrix $P_{U_n}(A)$ in M_{U_n} such that

$$\|A - X\|_2^2 \geq \left\|A - P_{U_n}(A)\right\|_2^2 \text{ for every } X \in M_{U_n}.$$

For each $A \in \mathbb{B}(\mathbb{H})$, consider the sequence of matrices $P_{U_n}(A_n)$ as the Frobenius optimal preconditioners of A_n. In the case A is the Toeplitz operator with continuous symbol, there are many good examples of matrix algebras such that the associated Frobenius optimal preconditioners are of low complexity and have faster rate of convergence.

3 Analytical Approach

The concepts of second-order relative spectra and quadratic projection method, which are almost synonyms of the other, were used in the spectral pollution problems and in determining the eigenvalues in the gaps by E.B. Davies, Levitin, Shagorodsky, etc. (see [5–7, 17]). In all these articles, the idea is to reduce the spectral approximation problems into the estimation of a particular function, related to the distance from the spectrum. This particular function is usually approximated by a sequence of functions related to the eigenvalues of truncations of the operator under concern.

First, we shall briefly mention the work done by E.B. Davies [6] and E.B. Davies and M. Plum [7], which is of great interest, where he considered functions which are related to the distance from the spectrum.

3.1 Distance from the Spectrum

In the paper published in 1998 [6], E.B. Davies considered the function F defined by

$$F(t) = \inf\left\{\frac{\|A(x) - tx\|}{\|x\|} : 0 \neq x \in \mathbb{L}\right\} \tag{3.1}$$

where $\mathbb{L}$ is a subspace of $\mathbb{H}$. Then he observed the following.

- F is Lipschitz continuous and satisfies $|F(s) - F(t)| \leq |s - t|$, for all $s, t \in \mathbb{R}$.
- $F(t) \geq d(t, \sigma(A)) = dist(t, \sigma(A))$.
- If $0 \leq F(t) \leq \delta$, then $\sigma(A) \cap [t - \delta, t + \delta] \neq \emptyset$.

From these observations, he obtained some bounds for the eigenvalues in the spectral gap of A, and found it useful in some concrete situations. For the efficient computation of the function F, he considered family of operators $N(s)$ on the given finite dimensional subspace $\mathbb{L}$, defined by

$$N(s) = A_{\mathbb{L}}^* A_{\mathbb{L}} - 2sPA_{\mathbb{L}} + s^2 I_{\mathbb{L}} \tag{3.2}$$

where P is the projection onto $\mathbb{L}$ and the notation $A_{\mathbb{L}}$ means A restricted to $\mathbb{L}$. The eigenvalues of these family of finite dimensional operators form sequence of real analytic functions (functions which map s to the eigenvalues of $N(s)$). He used these sequence to approximate the function F and thereby obtain information about the spectral properties of A. The main result is stated below, under the assumption that A is bounded.

Theorem 3.1 *Suppose $\{\mathbb{L}_n\}_{n=1}^{\infty}$ is an increasing sequence of closed subspaces of $\mathbb{H}$. If F_n is the functions associated with $\mathbb{L}_n$ according to (3.1), then F_n decreases monotonically and converge locally uniformly to $d(., \sigma(A))$. In particular, $s \in \sigma(A)$ if and only if*

$$\lim_{n\to\infty} F_n(s) = 0.$$

In the article on spectral pollution [7] in 2004, the above method was linked with various techniques due to Lehmann [16], Behnke et al. [3], Zimmerman et al. [22]. The problem of spurious eigenvalues in a spectral gap was addressed by considering the following function.

$$F(t) = \inf\{\|A(x) - tx\| : x \in \mathbb{L}, \|x\| = 1\}$$

If we define $F_n(t) = \inf\{\|A(x) - tx\| : x \in \mathbb{L}_n, \|x\| = 1\}$, then the following results shall be obtained.

- Given $\epsilon > 0$, there exists an N_ϵ such that $n \geq N_\epsilon$ implies

$$F(t) \leq F_n(t) \leq F(t) + \epsilon \text{ for all } t \in \mathbb{R}$$

- $\sigma(A) \cap [t - F_n(t), t + F_n(t)] \neq \emptyset$ for every $t \in \mathbb{R}$.

These observations were useful in obtaining some bounds for the eigenvalues between the bounds of essential spectrum. This was established with some numerical evidence in [7] for bounding eigenvalues for some particular operators.

Levitin and Shargorodsky considered the problem of spectral pollution in [17]. They suggested the usage of second-order relative spectra, to deal the problem. For the sake of completion, the definition is given below.

Definition 3.2 ([17]) Let $\mathbb{L}$ be a finite dimensional subspace of $\mathbb{H}$. A complex number z is said to belong to the second-order spectrum $\sigma_2(A, \mathbb{L})$ of A relative to $\mathbb{L}$ if there exists a nonzero u in $\mathbb{L}$ such that

$$\langle (A - zI)u, (A - \bar{z}I)v \rangle = 0, \text{ for every } v \in \mathbb{L}$$

They proved that the second-order relative spectrum intersects with every disk in the complex plane with diameter is an interval which intersect with the spectrum of A (Lemma 5.2 of [17]). They also provided some numerical results in case of some Multiplication and Differential operators, which indicated the effectiveness of second-order relative spectra in avoiding the spectral pollution. In [5], Boulton and Levitin used the quadratic projection method to avoid spectral pollution in the case of some particular Schrodinger operators.

3.2 Distance from the Essential Spectrum

To predict the existence of a gap in the essential spectrum, we need to know whether a number λ in (ν, μ) belongs to the spectrum or not. If it is not a spectral value, then there exists an open interval between (ν, μ) as a part of the compliment of the spectrum, since the compliment is an open set. We observe that the spectral gap prediction is possible by computing values of the following function.

Definition 3.3 Define the nonnegative valued function f on the real line $\mathbb{R}$ as follows.

$$f(\lambda) = \nu_\lambda = \inf \sigma_e((A - \lambda I)^2).$$

The primary observation is that we can predict the existence of a gap inside the essential spectrum by evaluating the function and checking whether it attains a nonzero value. The nonzero values of this function give the indication of spectral gaps.

Theorem 3.4 *The number λ in the interval (ν, μ) is in the gap if and only if $f(\lambda) > 0$. Also one end point of the gap will be $\lambda \pm \sqrt{f(\lambda)}$.*

The advantage of considering $f(\lambda)$ is that, it is the lower bound of the essential spectrum of the operator $(A - \lambda I)^2$, which we can compute using the finite dimensional truncations with the help of Theorem 2.18. So the computation of $f(\lambda)$, for each λ, is possible. This enables us to predict the gap using truncations. Also here we are able to compute one end point of a gap. The other end point is possible to compute by Theorem 2.3 of [18], which is stated below.

Theorem 3.5 ([18]) *Let A be a bounded self-adjoint operator and $\sigma_e(A) = [a, b] \bigcup [c, d]$, where $a < b < c < d$. Assume that b is known and not an accumulation point of the discrete spectra of A. Then c can be computed by truncation method.*

Coming back to the Arveson's class, we observe that the essential points and hence the essential spectrum is fully determined by the zeros of the function in the Definition 3.3.

Corollary 3.6 *If A is a bounded self-adjoint operator in the Arveson's class, then λ is an essential point if and only if $f(\lambda) = 0$.*

When one wishes to apply the above results to determine the gaps in the essential spectrum of a particular operator, one has to face the following problems. To check for each λ in (ν, μ), is a difficult task from the computational point of view. Also taking truncations of the square of the operator may lead to difficulty. Note that $(P_n A P_n)^2$ and $P_n A^2 P_n$ are entirely different. So we may have to do more computations to handle the problem.

Another problem is the rate of convergence and estimation of the remainder term. For each λ in (ν, μ), the value of the function $f(\lambda)$ has to be computed. This computation involves truncation of the operator $(A - \lambda I)^2$ and the limiting process of sequence of eigenvalues of each truncation. The rate of convergence of these approximations and the remainder estimate are the questions of interest.

Below, the function $f(.)$ is approximated by a double sequence of functions, which arise from the eigenvalues of truncations of operators.

Theorem 3.7 ([14]) *Let $f_{n,k}$ be the sequence of functions defined by $f_{n,k}(\lambda) = \lambda_{n+1-k}\left(P_n (A - \lambda I)^2 P_n\right)$. Then $f(.)$ is the uniform limit of a subsequence of $\{f_{n,k}(.)\}$ on all compact subsets of the real line.*

The following result makes the computation of $f(\lambda)$ much easier for a particular class of operators. When the operator is truncated first and square the truncation rather than truncating the square of the operator, the difficulty of squaring a bounded operator is reduced. The computation needs only to square the finite matrices.

Theorem 3.8 ([14]) *If $\|P_n A - A P_n\| \to 0$ as $n \to \infty$, then*

$$\lim_{k\to\infty} \lim_{n\to\infty} \lambda_{n+1-k}\left(P_n (A - \lambda I)^2 P_n\right) = \lim_{k\to\infty} \lim_{n\to\infty} \lambda_{n+1-k}\left(P_n (A - \lambda I) P_n\right)^2.$$

Remark 3.9 The function $f(.)$ that is considered here is directly related to the distance from the essential spectrum, while Davies' function was related with the distance from the spectrum. Here the approximation results in [4], especially Theorem 2.18 are used to approximate the function. But it is still not known to us whether these results are useful from a computational point of view. The methods due to Davies et al. were applied in the case of some Schrodinger operators with a particular kind of potentials in [5, 17]. We hope that a combined use of both methods may give a better understanding of the spectrum.

4 Concluding Remarks and Further Problems

The goal of such developments is to use the finite dimensional techniques into the spectral analysis of bounded self-adjoint operators on infinite dimensional Hilbert spaces. This also leads to a large number of open problems of different flavors. We shall quote some of them here.

- The numerical algorithms have to be developed to approximate spectrum and essential spectrum using the eigenvalue sequence of truncations, with emphasis on the computational feasibility.
- The random versions of the spectral approximation problems are another area to be investigated. The related work is already under progress in [14].
- The use of preconditioners has its origin in the numerical linear algebra literature, especially in the case of Toeplitz operators. One can expect good estimates on such concrete examples.
- The unbounded operators shall be considered and the approximation techniques have to be developed.

Acknowledgments The author wishes to thank Prof. M.N.N. Namboodiri for fruitful discussions. Also, thanks to National Board for Higher Mathematics (NBHM) for financial support and to the authority of Department of Mathematics, CUSAT, Cochin, for the hospitality during the International Conference on Semi groups Algebras and Operator Theory (ICSAOT-2014), 26–28 February 2014.

References

1. Arveson, W.B.: C^*- algebras and numerical linear algebra. J. Funct. Anal. **122**(2), 333–360 (1994)
2. Bhatia, R.: Matrix analysis. Graduate Text In Mathematics. Springer, New York (1997)
3. Behnke, H., Goerisch, F.: Inclusions for eigenvalues of self-adjoint problems. In: Herzberger, J. (ed.) Topics in Validated Computation, pp. 277–322. Elsevier, Amsterdam (1994)
4. Böttcher, A., Chithra, A.V., Namboodiri, M.N.N.: Approximation of approximation numbers by truncation. Integral Equ. Oper. Theory **39**, 387–395 (2001)
5. Boulton, L., Levitin, M.: On Approximation of the Eigenvalues of Perturbed Periodic Schrodinger Operators (2007). arxiv:math/0702420v1
6. Davies, E.B.: Spectral Enclosures and complex resonances for self-adjoint Operators. LMS J. Comput. Math. **1**, 42–74 (1998)
7. Davies, E.B., Plum, M.: Spectral Pollution. IMA J. Numer. Anal. **24**, 417–438 (2004)
8. Flaschka, H.: Discrete and periodic illustrations of some aspects of the inverse method. Lect. Notes Phys. **38**, 441–466 (1975)
9. Gohberg, I., Goldberg, S., Kaashoek, M.A.: Classes of Linear Operators, vol. I. Birkhguser Verlag, Basel (1990)
10. Golinskii, L., Kumar, K., Namboodiri, M.N.N., Serra-Capizzano, S.: A note on a discrete version of Borg's Theorem via Toeplitz-Laurent operators with matrix-valued symbols. Boll. Unione Mat. Ital. (9) **6**(1), 205–218 (2013)
11. Hansen, A.C.: On the approximation of spectra of linear operators on Hilbert spaces. J. Funct. Anal. **254**, 2092–2126 (2008)

12. Kumar, K., Namboodiri, M.N.N., Serra-Capizzano, S.: Preconditioners and Korovkin type Theorems for infinite dimensional bounded linear operators via completely positive maps. Studia Math. **218**(2), 95–118 (2013)
13. Kumar, K., Namboodiri, M.N.N., Serra-Capizzano, S.: Perturbation of operators and approximation of spectrum. Proc. Indian Acad. Sci. Math. Sci. **124**(2), 205–224 (2014)
14. Kumar, K.: Truncation method for random bounded self-adjoint operators. Banach J. Math. Anal. **9**(3), 98–113 (2015)
15. Kumar, K.: Preconditioners in spectral approximation. In: Proceedings of the UGC Sponsored National Seminar on Mathematical Analysis and Algebra (MAA 2013), held at PRNSS college, Mattanur, 05–06 September 2013
16. Lehmann, N.J.: Optimale Eigenwerteinschliessungen. Numer. Math. **5**, 246–272 (1963)
17. Levitin, M., Shargorodsky, E.: Spectral pollution and second-order relative spectra for self-adjoint operators IMA J. Numer. Anal. **24**, 393–416 (2004)
18. Namboodiri, M.N.N.: Truncation method for operators with disconnected essential spectrum. Proc. Indian Acad. Sci. (MathSci) **112**, 189–193 (2002)
19. Namboodiri, M.N.N.: Theory of spectral gaps- a short survey. J. Anal. **12**, 1–8 (2005)
20. Serra-Capizzano, S.: A Korovkin-type theory for finite Toeplitz operators via matrix algebras. Numerische Mathematik, vol. 82, pp. 117–142. Springer (1999)
21. Szegö, G.: Beitrge zur Theorie der Toeplitzschen Formen. Math. Z. **6**, 167–202 (1920)
22. Zimmerman, S., Mertins, U.: Variational bounds to eigenvalues of self-adjoint eigenvalue problems with arbitrary spectrum, Zeit. fr Anal. und ihre Anwendungen. J. Anal. Appl. **14**, 327–345 (1995)

On k-Minimal and k-Maximal Operator Space Structures

P. Vinod Kumar and M.S. Balasubramani

Abstract Let X be a Banach space and k be a positive integer. Suppose that we have matrix norms on $M_2(X)$, $M_3(X)$,..., $M_k(X)$ that satisfy Ruan's axioms. Then it is always possible to define matrix norms on $M_{k+1}(X)$, $M_{k+2}(X)$, ..., such that X becomes an operator space. As in the case of minimal and maximal operator spaces, here also we have a minimal and a maximal way to complete the sequence of matrix norms on X and this leads to ***k-minimal*** and ***k-maximal*** operator space structures on X. These spaces were first noticed by Junge [10] and more generally studied by Lehner [11]. Recently, the relationship of k-minimal and k-maximal operator space structures to norms that have been used in quantum information theory have been investigated by Johnston et al. [9]. We discuss some properties of these operator space structures.

Keywords Operator spaces · Completely bounded mappings · Minimal and maximal operator spaces · k-minimal operator space · k-maximal operator space

AMS Mathematics Subject Classification (2000) No 46L07 · 47L25

1 Introduction

The theory of operator spaces is a fairly new and rapidly developing branch of functional analysis and it can be regarded as the quantization of the theory of Banach spaces. The observables of classical mechanics are scalar valued functions, and in Heisenberg's theory of Quantum Mechanics these are replaced with infinite matrices

P. Vinod Kumar (✉)
Department of Mathematics, Thunchan Memorial Government College,
Tirur, Kerala, India
e-mail: vinodunical@gmail.com

M.S. Balasubramani
Department of Mathematics, University of Calicut,
Calicut University. P. O., Calicut, Kerala, India
e-mail: msbalaa@rediffmail.com

P.G. Romeo et al. (eds.), *Semigroups, Algebras and Operator Theory*,
Springer Proceedings in Mathematics & Statistics 142,
DOI 10.1007/978-81-322-2488-4_16

which can be regarded as operators on a Hilbert space. Through this quantization, noncommutativity comes into the picture and noncommutative or quantized versions of classical mathematical theories began to emerge as generalizations. A Banach space is a space of continuous functions on a compact Hausdorff space up to isometric isomorphism, where as an operator space is a space of operators on a Hilbert space, up to complete isometric isomorphism. So the operator spaces can be regarded as noncommutative normed spaces and their theory can be viewed as noncommutative functional analysis or quantized functional analysis. The main difference between the category of Banach spaces and that of operator spaces lies not in the spaces, but in the morphisms. A Banach space can be viewed as an operator space in a natural way, where as operator spaces are Banach spaces but here we simultaneously consider spaces of matrices associated with it. The appropriate morphisms for operator spaces are the completely bounded maps, that is, linear maps which induces uniformly bounded linear mappings on the associated spaces of matrices.

Given a Hilbert space $\mathcal{H}$, let $B(\mathcal{H})$ be the space of all bounded linear operators on $\mathcal{H}$. If X is a linear space, for each natural number n, $M_n(X)$ denotes the space of all $n \times n$ matrices over X and is called the nth *matrix level* of X.

If X and Y are linear spaces and $\varphi : X \to Y$ is a linear map, we can have natural amplifications (linear) $\varphi^{(n)} : M_n(X) \to M_n(Y)$, for each $n \in \mathbb{N}$, given by $[x_{ij}] \to [\varphi(x_{ij})]$, where $[x_{ij}] \in M_n(X)$. Suppose that each of the matrix levels of X and Y has given norms $\|.\|_{M_n(X)}$ and $\|.\|_{M_n(Y)}$, respectively.

A map $\varphi : X \to Y$ is said to be *k-bounded*, if $\varphi^{(k)} : M_k(X) \to M_k(Y)$ is bounded. The map φ is *completely bounded* if $\sup\{\|\varphi^{(n)}\| \mid n \in \mathbb{N}\} < \infty$ and we set $\|\varphi\|_{cb} = \sup\{\|\varphi^{(n)}\| \mid n \in \mathbb{N}\} = \sup\{\|[\varphi(x_{ij})]\|_{M_n(Y)} \mid \|[x_{ij}]\|_{M_n(X)} \leq 1, n \in \mathbb{N}\}$.

The map φ is said to be a *complete isometry* if each map $\varphi^{(n)} : M_n(X) \to M_n(Y)$ is an isometry. If φ is a complete isometry, then $\|\varphi\|_{cb} = 1$. The map φ is said to be *completely contractive* if $\|\varphi\|_{cb} \leq 1$. If $\varphi : X \to Y$ is a completely bounded linear bijection and if its inverse is also completely bounded, then φ is said to be a *complete isomorphism.*We denote the closed unit ball $\{x \in X \mid \|x\| \leq 1\}$ of X as $Ball$(X).

A *(concrete) operator space* X is a closed linear subspace of $B(\mathcal{H})$. Here, in each matrix level $M_n(X)$, we have a norm $\|.\|_n$, induced by the inclusion $M_n(X) \subset M_n(B(\mathcal{H}))$, where the norm in $M_n(B(\mathcal{H}))$ is given by the natural identification $M_n(B(\mathcal{H})) \approx B(\mathcal{H}^n)$, where $\mathcal{H}^n$ denotes the Hilbert space direct sum of n copies of $\mathcal{H}$.

An *abstract operator space,*or simply an *operator space* is a pair $(X, \{\|.\|_n\}_{n\in\mathbb{N}})$ consisting of a linear space X and a complete norm $\|.\|_n$ on $M_n(X)$ for every $n \in \mathbb{N}$, such that there exists a linear complete isometry $\varphi : X \to B(\mathcal{H})$ for some Hilbert space $\mathcal{H}$. The sequence of matrix norms $\{\|.\|_n\}_{n\in\mathbb{N}}$ is called an *operator space structure* on the linear space X. An operator space structure on a normed space $(X, \|.\|)$ will usually mean a sequence of matrix norms $\{\|.\|_n\}_{n\in\mathbb{N}}$ as above, but with $\|.\|_1 = \|.\|$. Two abstract operator spaces are considered to be the same if there is a complete isometric isomorphism from X to Y. In that case, we write $X \approx Y$ *completely isometrically.* In 1988, Z-J. Ruan characterized abstract operator spaces in terms of two properties of matrix norms.

Theorem 1.1 (Ruan [19]) *Suppose that X is a linear space and that for each $n \in \mathbb{N}$, we are given a norm $\|.\|_n$ on $M_n(X)$. Then X is completely isometrically isomorphic to a linear subspace of $B(\mathcal{H})$ for some Hilbert space $\mathcal{H}$ if and only if*

(R1) $\|\alpha x \beta\|_n \leq \|\alpha\| \, \|x\|_n \, \|\beta\|$ *for all* $\alpha, \beta \in M_n$ *and for all* $x \in M_n(X)$, *and*
(R2) $\|x \oplus y\|_{m+n} = max\{\|x\|_m, \|y\|_n\}$ *for all* $x \in M_m(X)$, *and for all* $y \in M_n(X)$,

where $x \oplus y$ denotes the matrix $\begin{bmatrix} x & 0 \\ 0 & y \end{bmatrix}$ in $M_{m+n}(X)$ where 0 stands for zero matrices of appropriate orders.

The conditions (R1) and (R2) are usually known as Ruan's axioms.

Thus we see that an abstract operator space is simply a Banach space X together with a sequence of matrix norms that satisfies the Ruan's axioms. If X is a concrete operator space, then in each matrix level $M_n(X)$, we have a norm $\|.\|_n$, induced by the inclusion $M_n(X) \subset M_n(B(\mathcal{H}))$, where the norm in $M_n(B(\mathcal{H}))$ is given by the natural identification $M_n(B(\mathcal{H})) \approx B(\mathcal{H}^n)$. Also, these induced matrix norms satisfy Ruan's axioms. Thus, every concrete operator space can be regarded as an abstract operator space. Ruan's theorem allows us to view an operator space in an abstract way free of any concrete representation on a Hilbert space and so, we no longer distinguish between concrete and abstract operator spaces.

Many concepts from Banach space theory can be formulated in the settings of this quantized theory of Banach spaces, and this provided important generalizations of many results. Studies have shown that this theory gives a more general setup to study the structure of operator algebras. Certain invariants of operator algebras like injectivity, exactness, and local reflexivity can be understood in a better way as the properties of their underlying operator space structures. Also, the matricial orderings and sequence of matrix norms played an important role in the algebraic classification of operator algebras. Operator space theory, thus, serves as a bridge between the theory of Banach spaces and that of operator algebras. More information about operator spaces and completely bounded mappings may be found in the papers [4, 5, 14, 18] or in the recent monographs [3, 6, 16, 17].

2 Operator Space Structures on Banach Spaces

Given a Banach space X, there are many operator space structures possible on X, which all have X as their first matrix level. Blecher and Paulsen [1] observed that the set of all operator space structures admissible on a given Banach space X admits a minimal and maximal element namely $Min(X)$ and $Max(X)$, which represent, respectively, the smallest and the largest operator space structures admissible on X.

These structures were further investigated by Paulsen in [14, 15]. Much work has been done in understanding these operator space structures on a Banach space X, and these studies have played a vital role in the theory of operator spaces and in the theory of C*-algebras.

Let X be a Banach space and X^* be its dual space. Let $K = Ball(X^*)$ be the unit closed ball of the dual space of X with its weak* topology. Then the canonical embedding $J : X \to C(K)$, defined by $J(x)(f) = f(x), x \in X$ and $f \in K$ is a linear isometry. Since by Gelfand–Naimark theorem, subspaces of C*-algebras are operator spaces, this identification of X induces matrix norms on $M_n(X)$ that makes X an operator space. The matrix norms on X are given by

$$\left\|[x_{ij}]\right\|_n = \sup\{\left\|[f(x_{ij})]\right\| \mid f \in K\}$$

for all $[x_{ij}] \in M_n(X)$ and for all $n \in \mathbb{N}$.

Just as Banach spaces may be regarded as the subspaces of commutative C*-algebras, by Ruan's theorem, operator spaces may be viewed as the subspaces of $B(\mathcal{H})$, or by Gelfand–Naimark theorem, they are exactly the subspaces of general C*-algebras. So we can regard operator spaces as noncommutative Banach spaces.

The above-defined operator space structure on X is called the minimal operator space structure on X, and we denote this operator space as $Min(X)$. For $[x_{ij}] \in M_n(X)$, we write $\left\|[x_{ij}]\right\|_{Min(X)}$ to denote its norm as an element of $M_n(Min(X))$. This minimal quantization of a normed space is characterized by the property that for any arbitrary operator space Y and for any bounded linear map $\varphi : Y \to Min(X)$ is completely bounded and satisfies $\|\varphi : Y \to Min(X)\|_{cb} = \|\varphi : Y \to X\|$. Thus, if X and Y are Banach spaces and $\varphi \in B(X, Y)$, then φ is completely bounded and $\|\varphi\|_{cb} = \|\varphi\|$, when considered it as a map $X \to Min(Y)$.

If X is an operator space, for any $x \in M_n(X)$, there exists a complete contraction [6], $\varphi : X \to M_n$ such that $\|\varphi_n(x)\| = \|x\|$. Therefore, for $[x_{ij}] \in M_n(X)$, we have $\left\|[x_{ij}]\right\|_n = \sup\{\left\|f_n([x_{ij}])\right\| \mid f \in Ball(CB(X, M_n))\}$. This shows that $Min(X)$ is the smallest operator space structure on X. We say that an operator space X is minimal if $Min(X) = X$. An operator space is minimal if and only if it is completely isomorphic to a subspace of a commutative C*-algebra [6].

If X is a Banach space, there is a maximal way to consider it as an operator space. The matrix norms given by

$$\left\|[x_{ij}]\right\|_n = \sup\{\left\|[\varphi(x_{ij})]\right\| \mid \varphi \in Ball(B(X, Y))\}$$

where the supremum is taken over all operator spaces Y and all linear maps $\varphi \in Ball(B(X, Y))$, makes X an operator space. We denote this operator space as $Max(X)$ and is called the maximal operator space structure on X. For $[x_{ij}] \in M_n(X)$, we write $\left\|[x_{ij}]\right\|_{Max(X)}$ to denote its norm as an element of $M_n(Max(X))$. We say that an operator space X is maximal if $Max(X) = X$. By Ruan's Theorem 1.1, we also have

$$\left\|[x_{ij}]\right\|_{Max(X)} = \sup\{\left\|[\varphi(x_{ij})]\right\| \mid \varphi \in Ball(B(X, B(\mathcal{H})))\}$$

where the supremum is taken over all Hilbert spaces $\mathcal{H}$ and all linear maps $\varphi \in Ball(B(X, B(\mathcal{H})))$. By the definition of $Max(X)$, any operator space structure that we can put on X, must be smaller than $Max(X)$.

This maximal quantization of a normed space is characterized by the property that for any arbitrary operator space Y and for any bounded linear map $\varphi : Max(X) \to Y$ is completely bounded and satisfies $\|\varphi : Max(X) \to Y\|_{cb} = \|\varphi : X \to Y\|$. Thus, if X and Y are Banach spaces and $\varphi \in B(X, Y)$, then φ is completely bounded and $\|\varphi\|_{cb} = \|\varphi\|$, when considered it as a map $Max(X) \to Y$. If X is any operator space, then the identity map on X defines completely contractive maps $Max(X) \to X \to Min(X)$.

For any Banach space X, we have the duality relations $Min(X)^* = Max(X^*)$ and $Max(X)^* = Min(X^*)$ completely isometrically [2].

By Hahn–Banach theorem, any subspace of a minimal operator space is again minimal, but the quotient of a minimal space need not be minimal. The subspace of a maximal space need not be maximal and such spaces are called *submaximal* spaces. But quotient spaces inherits the maximality. Subspace structure of various maximal operator spaces were studied in [12]. In [20], the notion of hereditarily maximal spaces is introduced. Hereditarily maximal spaces determine a subclass of maximal operator spaces with the property that the operator space structure induced on any subspace coincides with the maximal operator space structure on that subspace. An operator space X is homogeneous if each bounded linear operator φ on X is completely bounded with $\|\varphi\|_{cb} = \|\varphi\|$ [17]. The spaces $Min(X)$ and $Max(X)$ are homogeneous, but in general, submaximal spaces need not be homogeneous.

3 k-Minimal and k-Maximal Operator Spaces

We now focus on some generalizations of minimal and maximal operator space structure on a Banach space X. Let X be a Banach space, and k be a positive integer. Suppose that we have matrix norms on $M_2(X), M_3(X), \ldots, M_k(X)$ that satisfy Ruan's axioms. Then it is always possible to define matrix norms on $M_{k+1}(X)$, $M_{k+2}(X), \ldots,$ such that X becomes an operator space. As in the case of minimal and maximal operator spaces, here also we have a minimal and a maximal way to complete the sequence of matrix norms on X and this leads to *k-minimal* and *k-maximal* operator space structures on X. These spaces were first noticed by Junge [10] and more generally studied by Lehner [11]. Recently, the relationship of k-minimal and k-maximal operator space structures to norms that have been used in quantum information theory [7, 8] have been investigated by Johnston et al. [9].

If we define the matrix norms on $M_n(X)$ for $n > k$ as the matrix norms in $M_n(Min(X))$, the resulting operator space is called the *k-minimal operator space* and is denoted by $Min^k(X)$. An operator space X is said to be *k-minimal* if $Min^k(X) = X$.

Similarly, if we set the norms on $M_n(X)$ for $n > k$ as the matrix norms in $M_n(Max(X))$, the resulting operator space is called the *k-maximal operator space* and is denoted by $Max^k(X)$. An operator space X is said to be *k-maximal* if $Max^k(X) = X$.

If $(X, \{\|.\|_{M_n(X)}\})$ is any operator space, the matrix norms in $Min^k(X)$ and $Max^k(X)$ are explicitly given ([9, 13]) as follows:

$$\|[x_{ij}]\|_{M_n(Min^k(X))} = \sup\{\|[\varphi(x_{ij})]\| \mid \varphi : X \to M_k, \|\varphi\|_{cb} \leq 1\},$$

and

$$\|[x_{ij}]\|_{M_n(Max^k(X))} = \sup\{\|[\varphi(x_{ij})]\| \mid \varphi : X \to B(\mathcal{H}), \|\varphi^{(k)}\| \leq 1, \text{ all Hilbert spaces } \mathcal{H}\}$$

If $\{\|.\|'_{M_n(X)}\}$ be any operator space structure on X, such that for $1 \leq n \leq k$, $\|[x_{ij}]\|'_{M_n(X)} = \|[x_{ij}]\|_{M_n(X)}$, then

$$\|[x_{ij}]\|_{M_n(Min^k(X))} \leq \|[x_{ij}]\|'_{M_n(X)} \leq \|[x_{ij}]\|_{M_n(Max^k(X))}$$

for all $[x_{ij}] \in M_n(X)$ and for all $n \in \mathbb{N}$.

From the definitions of minimal and maximal operator spaces, we see that, when $k = 1$, the k-minimal and the k-maximal operator space structures on X coincides with the minimal and the maximal operator space structures on X, respectively.

The notions of k-minimal and k-maximal operator spaces help us to obtain several different operator space structures on a given operator space X. For instance, for any $k \in \mathbb{N}$, the space $Min^k(Max^{k-1}(X))$ is the space whose matrix norms up to the $(k-1)$th level are the same as those of X, on the kth level, the norms in $Max(X)$, and from $(k+1)$th level onward, are the matrix norms from $Min(X)$.

Remark 3.1 Let X be an operator space. It can be noted that the formal identity maps $Max^k(X) \to X \to Min^k(X)$ are completely contractive. Also, for $n \leq k$, the formal identity maps $id : M_n(Min^k(X)) \to M_n(X) \to M_n(Max^k(X))$ are isometries. From the definition, it follows that for $[x_{ij}] \in M_n(X)$, the sequence $\{\|[x_{ij}]\|_{M_n(Min^k(X))}\}$ increases to $\|[x_{ij}]\|_{M_n(X)}$ and the sequence $\{\|[x_{ij}]\|_{M_n(Max^k(X))}\}$ decreases to $\|[x_{ij}]\|_{M_n(X)}$ as $k \to \infty$.

The following duality relations [13] hold:
$(Min^k(X))^* \cong Max^k(X^*)$ and $(Max^k(X))^* \cong Min^k(X^*)$ completely isometrically.

4 Universal Properties of k-Minimal and k-Maximal Spaces

From the definitions of k-minimal and k-maximal spaces, the following observation is straightaway.

Proposition 4.1 *Let X be an operator space and $k, h \in \mathbb{N}$. Then the formal identity mapping, $id : Min^k(X) \to Min^h(X)$ is completely contractive whenever $k \geq h$, and $id : Max^k(X) \to Max^h(X)$ is completely contractive whenever $k \leq h$.*

We now show that k-minimal and k-maximal operator spaces are characterized up to complete isometric isomorphism in terms of some universal properties. A part of these characterizations can be found in [13] where these are described in a slightly different terminology.

Theorem 4.2 *An operator space Y is a k-minimal operator space up to complete isometric isomorphism if and only if for any operator space X and for any bounded linear map $\varphi : X \to Y$, we have $\|\varphi : X \to Y\|_{cb} = \|\varphi^{(k)}\|$.*

Proof Assume that $Y = Min^k(Y)$. We have,

$$\begin{aligned}
\|\varphi : X \to Y\|_{cb} &= \left\|\varphi : X \to Min^k(Y)\right\|_{cb} \\
&= \sup\{\left\|[\varphi(x_{ij})]\right\|_{M_n(Min^k(Y))} \mid \left\|[x_{ij}]\right\|_{M_n(X)} \leq 1, n \in \mathbb{N}\} \\
&= \sup\{\sup\{\left\|[\psi(\varphi(x_{ij}))]\right\| \mid \psi : Y \to M_k, \|\psi\|_{cb} \leq 1\}, \\
&\qquad\qquad \left\|[x_{ij}]\right\|_{M_n(X)} \leq 1, n \in \mathbb{N}\} \\
&= \sup\{\sup\{\left\|(\psi \circ \varphi)^{(n)}([x_{ij}])\right\| \mid \left\|[x_{ij}]\right\|_{M_n(X)} \leq 1, n \in \mathbb{N}\}, \\
&\qquad\qquad \psi : Y \to M_k, \|\psi\|_{cb} \leq 1\} \\
&= \sup\{\|\psi \circ \varphi\|_{cb} \mid \psi : Y \to M_k, \|\psi\|_{cb} \leq 1\} \qquad (1)
\end{aligned}$$

Now, $\psi \circ \varphi : X \to M_k$, by [18], $\|\psi \circ \varphi\|_{cb} = \left\|(\psi \circ \varphi)^{(k)}\right\| \leq \left\|\psi^{(k)}\right\| \left\|\varphi^{(k)}\right\|$. Thus, from the above equation (1), $\|\varphi : X \to Y\|_{cb} \leq \|\varphi^{(k)}\|$.
Thus, $\|\varphi : X \to Y\|_{cb} = \|\varphi^{(k)}\|$.

For the converse, take $X = Min^k(Y)$ and $\varphi = id$, the identity mapping. Then by assumption, $\|id : Min^k(Y) \to Y\|_{cb} = \|id^{(k)}\|$.

Since, $\|[x_{ij}]\|_{M_n(Y)} = \|[x_{ij}]\|_{M_n(Min^k(Y))}$ for $1 \leq n \leq k$, $\|id^{(k)}\| = 1$.

Also, $\|id^{-1} : Y \to Min^k(Y)\|_{cb} = \|id : Y \to Min^k(Y)\|_{cb} \leq 1$. Thus, id is a complete isometric isomorphism. □

Theorem 4.3 *An operator space X is a k-maximal operator space up to complete isometric isomorphism if and only if for any operator space Y and for any bounded linear map $\varphi : X \to Y$, we have $\|\varphi : X \to Y\|_{cb} = \|\varphi^{(k)}\|$.*

Proof Assume that $X = Max^k(X)$. Let $\varphi : X \to Y$ be any bounded linear map. Set $\upsilon = \dfrac{\varphi}{\|\varphi^{(k)}\|}$. Then, $\upsilon : X \to Y \subset B(\mathcal{H})$ is bounded and $\|\upsilon^{(k)}\| \leq 1$. Therefore, by the definition of matrix norms in $Max^k(X)$, $\|[\upsilon(x_{ij})]\|_{M_n(Y)} \leq \|[x_{ij}]\|_{M_n(Max^k(X))}$, for every $[x_{ij}] \in M_n(X)$ and for all $n \in \mathbb{N}$. Thus, $\|\upsilon\|_{cb} \leq 1$, implies that $\|\varphi\|_{cb} \leq \|\varphi^{(k)}\|$. Hence follows the desired result.

For the converse, take $Y = Max^k(Y)$ and $\varphi = id$, the identity mapping. Then by assumption, $\|id : X \to Max^k(X)\|_{cb} = \|id^{(k)}\| = 1$.

Also, $\|id^{-1} : Max^k(X) \to X\|_{cb} \leq 1$. Thus, id is a complete isometric isomorphism. □

Corollary 4.4 *Let X and Y be operator spaces, $k \in \mathbb{N}$, and $\varphi : X \to Y$ be a bounded linear map. Then,*

(i) φ is k-bounded if and only if $\varphi : Max^k(X) \to Y$ is completely bounded.
(ii) φ is k-bounded if and only if $\varphi : X \to Min^k(Y)$ is completely bounded.

Since a subspace of a maximal operator space need not be maximal, a subspace of a k-maximal space need not be k-maximal. But, subspaces of k-minimal spaces are again k-minimal.

Theorem 4.5 *If Y is a subspace of the operator space X, then $Min^k(Y)$ is a subspace of $Min^k(X)$.*

Proof We have to prove that $\|[y_{ij}]\|_{M_n(Min^k(Y))} = \|[y_{ij}]\|_{M_n(Min^k(X))}$, $\forall [y_{ij}] \in M_n(Y)$ and $\forall n \in \mathbb{N}$. By definition,

$$\|[y_{ij}]\|_{M_n(Min^k(Y))} = \sup\{\|[\varphi(y_{ij})]\| \mid \varphi : Y \to M_k, \|\varphi\|_{cb} \leq 1\},$$

and

$$\|[y_{ij}]\|_{M_n(Min^k(X))} = \sup\{\|[\tilde{\varphi}(x_{ij})]\| \mid \tilde{\varphi} : X \to M_k, \|\tilde{\varphi}\|_{cb} \leq 1\}.$$

Now, if $\tilde{\varphi} : X \to M_k$ is a complete contraction, then $\tilde{\varphi}|_Y : Y \to M_k$ is also a complete contraction.

On the other hand, since M_k is injective, any complete contraction $\varphi : Y \to M_k$ extends to a complete contraction $\tilde{\varphi} : X \to M_k$ such that $\|\tilde{\varphi}\|_{cb} = \|\varphi\|_{cb}$. Hence, both the matrix norms are the same. □

We know that for any Banach space X, the operator spaces $Min(X)$ and $Max(X)$ are homogeneous. Now we discuss the case of k-minimal and k-maximal spaces.

Theorem 4.6 *Let X be an operator space and $k \in \mathbb{N}$.*

(i) The spaces $Min^k(X)$ and $Max^k(X)$ are λ-homogeneous for some $\lambda > 0$.
(ii) If X is homogeneous, so are $Min^k(X)$ and $Max^k(X)$.
(iii) If X is homogeneous, then for any bounded linear operator φ on X, we have $\|\varphi : X \to Min^k(X)\|_{cb} = \|\varphi\|$ and $\|\varphi : Max^k(X) \to X\|_{cb} = \|\varphi\|$.

Proof We prove the results only for the k-minimal spaces. The other case will follow in a similar way.

Let $\varphi : Min^k(X) \to Min^k(X)$ be a bounded linear map. Then we have,

$\|\varphi : Min^k(X) \to Min^k(X)\|_{cb} = \|\varphi^{(k)}\|$. Now, by using the operator space matrix norm inequalities,

$$\begin{aligned}\|\varphi^{(k)}\| &= \sup\{\|[\varphi(x_{ij})]\|_{M_k(Min^k(X))} \mid \|[x_{ij}]\|_{M_k(Min^k(X))} \leq 1\} \\ &\leq \sup\{\Sigma_{i,j}\|\varphi(x_{ij})\| \mid \|[x_{ij}]\|_{M_k(Min^k(X))} \leq 1\} \\ &\leq k^2\|\varphi\|\end{aligned}$$

Thus, if X is any operator space, $Min^k(X)$ is λ-homogeneous for some $\lambda > 0$.

Now, to prove (ii), consider a bounded linear map $\varphi : Min^k(X) \to Min^k(X)$. Then by Theorem 4.2,

$$\begin{aligned}\|\varphi : Min^k(X) \to Min^k(X)\|_{cb} &= \|\varphi^{(k)} : M_k(Min^k(X)) \to M_k(Min^k(X))\| \\ &= \|\varphi^{(k)} : M_k(X) \to M_k(X)\| \\ &\leq \|\varphi : X \to X\|_{cb} \\ &= \|\varphi\|\end{aligned}$$

Thus, $Min^k(X)$ is homogeneous, if X is homogeneous.

For proving (iii), let $\varphi : X \to Min^k(X)$ be a bounded linear map. Let φ_0 denote the same map φ but, regarded as a map from $Min^k(X) \to Min^k(X)$. Then $\varphi = \varphi_0 \circ id$, where $id : X \to Min^k(X)$ is the identity map.

Since $id : X \to Min^k(X)$ is a complete contraction and by using (ii), we have $\|\varphi\|_{cb} = \|\varphi_0 \circ id\|_{cb} \leq \|id\|_{cb}\|\varphi_0\|_{cb} \leq \|\varphi_0\| = \|\varphi\|$. □

By using the universal properties of k-minimal and k-maximal spaces, we now obtain expressions for the cb-norm of the identity mappings $id : Min^k(X) \to X$ and $id : X \to Max^k(X)$.

Theorem 4.7 *Let X be an operator space and $k \in \mathbb{N}$.*
(i) The identity mapping $id : Min^k(X) \to X$ is completely bounded if and only if for every operator space Y and every k-bounded linear map $\varphi : Y \to X$ is completely bounded. Moreover, $\|id : Min^k(X) \to X\|_{cb} = sup\{\frac{\|\varphi\|_{cb}}{\|\varphi^{(k)}\|}\}$ where the supremum is taken over all k-bounded nonzero linear maps $\varphi : Y \to X$ and all operator spaces Y.
(ii) The identity mapping $id : X \to Max^k(X)$ is completely bounded if and only if for every operator space Y and every k-bounded linear map $\varphi : X \to Y$ is completely bounded. Moreover, $\|id : X \to Max^k(X)\|_{cb} = sup\{\frac{\|\varphi\|_{cb}}{\|\varphi^{(k)}\|}\}$ where the supremum is taken over all k-bounded nonzero linear maps $\varphi : X \to Y$ and all operator spaces Y.

Proof We prove only (i) and (ii) will follow in a similar way. Assume that the identity mapping $id : Min^k(X) \to X$ is completely bounded. Let $\varphi : Y \to X$ is k-bounded. Let $\widetilde{\varphi}$ be the same map as φ but with $Min^k(X)$ as the range. Then $\varphi = id \circ \widetilde{\varphi}$. Now by Theorem 4.2, $\|\widetilde{\varphi}\|_{cb} = \|\widetilde{\varphi}^{(k)}\| = \|\varphi^{(k)}\| < \infty$. Since φ is the composition of two completely bounded maps, it is completely bounded.

For proving the converse, take $Y = Min^k(X)$ and $\varphi = id$.

Now, $id^{(k)} = M_k(Min^k(X)) \to M_k(X)$ is an isometry, and by assumption, we see that $id : Min^k(X) \to X$ is completely bounded.

Since, $\varphi = id \circ \widetilde{\varphi}$, $\|\varphi\|_{cb} \leq \|\widetilde{\varphi}\|_{cb}\|id\|_{cb}$. But $\|\widetilde{\varphi}\|_{cb} = \|\widetilde{\varphi}^{(k)}\| = \|\varphi^{(k)}\|$, so that $\frac{\|\varphi\|_{cb}}{\|\varphi^{(k)}\|} \leq \|id\|_{cb}$. Since $id : Min^k(X) \to X$ is also a member of the right side collection, we get the desired equality. □

Theorem 4.8 *Let X be an operator space, and $k \in \mathbb{N}$. Then X is k-minimal (k-maximal) if and only if the bidual X^{**} is k-minimal (k-maximal).*

Proof Assume that $X = Min^k(X)$. Then by duality relations, $X^* = (Min^k(X))^* = Max^k(X^*)$, so that $X^{**} = (Max^k(X^*))^* = Min^k(X^{**})$. Thus X^{**} is k-minimal. Since $X \subset X^{**}$, if X^{**} is k-minimal, by Theorem 4.5, we see that X is k-minimal. The k-maximal case will follow in a similar way, where the reverse implication can be obtained by using the universal property. □

Acknowledgments The first author acknowledges the financial support by University Grants Commission of India, under Faculty Development Programme. The authors are grateful to Prof. Gilles Pisier and Prof. Eric Ricard for having many discussions on this subject.

References

1. Blecher, D., Paulsen, V.: Tensor products of operator spaces. J. Funct. Anal. **99**, 262–292 (1991)
2. Blecher, D.: The standard dual of an operator space. Pac. J. Math. **153**, 15–30 (1992)
3. Blecher, D., Le Merdy, C.: Operator Algebras and their Modules. An Operator Space Approach. London Mathematical Society Monographs. Oxford University Press, Oxford (2004)
4. Effros, E., Ruan, Z.-J.: A new approach to operator spaces. Can. Math. Bull. **34**, 329–337 (1991)
5. Effros, E., Ruan, Z.-J.: On the abstract characterizations of operator spaces. Proc. Am. Math. Soc. **119**, 579–584 (1993)
6. Effros, E., Ruan, Z.-J.: Operator Spaces. London Mathematical Society Monographs. New Series. Oxford University Press, Oxford (2000)
7. Johnston, N., Kribs, D.W.: A family of norms with applications in quantum information theory. J. Math. Phys. **51**, 082202 (2010)
8. Johnston, N., Kribs, D.W.: A family of norms with applications in quantum information theory—II. Quantum Inf. Comput. **11**, 104–123 (2011)
9. Johnston, N., Kribs, D.W., Paulsen, V., Pereira, R.: Minimal and maximal operator spaces and operator systems in entanglement theory. J. Funct. Anal. **260**, 2407–2423 (2011)
10. Junge, M.: Factorization theory for spaces of operators. Habilitationsschrift, Universitat Kiel (1996)
11. Lehner, F.: Mn espaces, sommes dunitaires et analyse harmonique sur le groupe libre. Ph.D. thesis, Universite Paris VI (1997)
12. Oikhberg, T.: Subspaces of maximal operator spaces. Integral Equ. Oper. Theory **48**, 81–102 (2004)
13. Oikhberg, T., Ricard, E.: Operator spaces with few completely bounded maps. Math. Ann. **328**, 229–259 (2004)
14. Paulsen, V.: Representation of function algebras, abstract operator spaces and Banach space geometry. J. Funct. Anal. **109**, 113–129 (1992)

15. Paulsen, V.: The maximal operator space of a normed space. Proc. Edinb. Math. Soc. **39**, 309–323 (1996)
16. Paulsen, V.: Completely Bounded maps and Operator Algebras. Cambridge University Press, Cambridge (2002)
17. Pisier, G.: Introduction to Operator Space Theory. Cambridge University Press, Cambridge (2003)
18. Smith, R.R.: Completely bounded maps between C^*-algebras. J. Lond. Math. Soc. **27**, 157–166 (1983)
19. Ruan, Z.-J.: Subspaces of C^*-algebras. J. Funct. Anal. **76**, 217–230 (1988)
20. Vinod Kumar, P., Balasubramani, M.S.: Submaximal operator space structures on Banach spaces. Oper. Mat. **7**(3), 723–732 (2013)

Erratum to: Spectral Approximation of Bounded Self-Adjoint Operators—A Short Survey

K. Kumar

Erratum to:
Chapter 'Spectral Approximation of Bounded Self-Adjoint Operators—A Short Survey' in: P.G. Romeo et al. (eds.), *Semigroups, Algebras and Operator Theory*, Springer Proceedings in Mathematics & Statistics 142, DOI 10.1007/978-81-322-2488-4_15

The original version of this article was inadvertently published with an incorrect abstract for chapter 15. The correct abstract appears here

Abstract A survey of the different techniques used to approximate the spectrum of bounded self-adjoint operators on separable Hilbert spaces, is presented here. Approximating an infinite dimensional operator by its finite dimensional truncations were useful to approximate the eigenvalues of compact operators. The lack of operator norm convergence makes it difficult in the case of non compact operators. In 1994, W.B. Arveson identified a class of operators for which the finite dimensional truncations are useful in the spectral approximation. The C^*-algebraic approach due to Arveson was a landmark in the theory of spectral approximation. Later, some progress was made with the crucial assumption; connectedness of the essential spectrum. The spectral pollution problems and spectral gap problems were also addressed by many mathematicians. The use of the quadratic projection

---

The online version of the original chapter can be found under
DOI 10.1007/978-81-322-2488-4_15

K. Kumar (✉)
Department of Mathematics, PSMO College, Tirurangadi 676306, Kerala, India
e-mail: kiranbalu36@gmail.com

P.G. Romeo et al. (eds.), *Semigroups, Algebras and Operator Theory*,
Springer Proceedings in Mathematics & Statistics 142,
DOI 10.1007/978-81-322-2488-4_17

method and second order relative spectra are also discussed in this article. Some of the recent results in the spectral gap prediction problems are explained here. Also, we try to modify the truncation method by using the notion of preconditioners and matrix convergence in the sense of eigenvalue clustering.